MEASURING GLOBAL TEMPERATURES

Temperature is probably the most influential of all climatic variables. Our only direct, quantitative knowledge of global temperatures (and indeed of any climate variable) comes from instruments invented and operated over the last 150 or so years. Crucial as these data are to our understanding of the climate, they are largely taken for granted, even by many of those using them, with little appreciation of how they came into being, how accurate they are, or how urgently they need to be improved for the future.

Measuring Global Temperatures will fill this gap by explaining how global temperatures are measured, how the data are analysed, what the potential errors are, and what needs to be done to improve temperature measurement in the future. After looking at how temperature measurements are made, Ian Strangeways shows how they are then processed to give average temperatures over the globe, explaining that global values may be misleading because they merge many different climates and seasons into one and so may hide vital details. Using some of the best datasets currently available, the book shows that temperature change is complex, and not reducible to one single graph or global truth: we need to look at the local scale for clues as to how and why the climate is changing.

This book will be of great interest to all meteorologists, climatologists and hydrologists, especially those concerned with climate change and global warming. It is written in accessible language with little mathematics, and will also appeal to students and anyone with an interest in weather and climate.

IAN STRANGEWAYS obtained his degree in electronic engineering, physics and mathematics from Bangor University, followed some years later by a PhD in meteorological instrumentation from Reading. After 24 years at the Institute of Hydrology as head of the Applied Physics Section, measuring the hydrometeorological environment, he became Director of TerraData, a consultancy in meteorological and hydrological data collection. He is the author of *Measuring the Natural Environment*, now in its second edition (2003, Cambridge University Press) and *Precipitation: Theory, Measurement and Distribution* (2007, Cambridge University Press). He has published many papers on the topic of data collection, climate measurement and instrumentation, and has written extensively for the Royal Meteorological Society's (RMetS) magazine *Weather*, for which he received the Gordon Manley Award in 2005. He is on the committee of the RMetS Special Interest Group on Meteorological Observing Systems, which he chaired for three years. He has travelled extensively to many remote areas of the world, advising on the monitoring of the weather and rivers, and has given talks at major international conferences. He has been an external lecturer on the subject at Newcastle University, and has also taught at the Open University and instructed students overseas. He is a Fellow of the RMetS and an Associate Member of the British Hydrological Society and the Institution of Civil Engineers.

MEASURING GLOBAL TEMPERATURES

Their Analysis and Interpretation

IAN STRANGEWAYS

CAMBRIDGE
UNIVERSITY PRESS

CAMBRIDGE UNIVERSITY PRESS
Cambridge, New York, Melbourne, Madrid, Cape Town, Singapore,
São Paulo, Delhi, Dubai, Tokyo

Cambridge University Press
The Edinburgh Building, Cambridge CB2 8RU, UK

Published in the United States of America by Cambridge University Press, New York

www.cambridge.org
Information on this title: www.cambridge.org/9780521898485

First published 2010

Printed in the United Kingdom at the University Press, Cambridge

A catalogue record for this publication is available from the British Library

Library of Congress Cataloguing in Publication data
Strangeways, Ian, 1932–
Measuring global temperatures : analysis and interpretation / Ian C. Strangeways.
p. cm.
Includes bibliographical references and index.
ISBN 978-0-521-89848-5 (hardback)
1. Atmospheric temperature–Observations. 2. Atmospheric temperature–Measurement.
3. Global temperature changes. I. Title.
QC901.A1S78 2009
551.5′250287–dc22 2009028741

ISBN 978-0-521-89848-5 Hardback

In writing this book, I have adopted the Royal Society's motto:

Nullius in verba

(*Nothing upon another's word*)

(Horace, Epistles I.i, 1.13–14)

Contents

Preface

In the course of my work measuring the weather and rivers, I have had the good fortune to visit many remote locations around the world that few people get to see. The purpose of these journeys was to advise on how the meteorological and hydrological environments should be measured for applications ranging from a country's water resources, through flood prediction and warning, agriculture and irrigation in deserts, plant diseases in the tropics, water for hydroelectric schemes, the transmission of measurements to remote bases via telephone, radio and satellite, and a wide variety of meteorological measurements for diverse applications. When not thus occupied I spend my time in laboratories in the UK developing new ways to measure the natural environment and writing up the accumulated knowledge gained in this way over four decades.

This work has taken me into the rainforests of Papua New Guinea, Indonesia and Malaysia, into the deserts of Libya and Oman, to the Cairngorm mountain-tops in winter, to hydroelectric generator halls inside the mountains of Honduras and to remote islands off the Antarctica peninsula, testing experimental instruments, and by small plane to grass airstrips in the Mato Grosso of the Brazilian interior, to India, Malaysia, Indonesia and to many other impressive locations. Such journeys are not without stress, for it was vital that what I did there had to work and I had to 'hit the ground running' despite jet lag and had to avoid getting ill despite some primitive living conditions.

Seeing at first hand the way in which different countries undertook the task of making their measurements, and the field conditions in diverse climates, expanded my view beyond the comfortable confines of laboratories and offices out into many corners of the globe where scientific standards vary greatly and where you are on your own without any help immediately to hand. Travel to these places has coloured my view of the world profoundly.

Out of these many journeys came my first book, *Measuring the Natural Environment*, and two years later its second edition, describing how meteorological and hydrological variables are measured. The next book was *Precipitation*, which concentrated on just that one topic in detail; on a history of our understanding of what it is, on the way it is measured and on what the measurements tell us about rainfall and snowfall around the globe. After that I felt it was time to turn my attention to what is now undoubtedly the most important of all climate variables, temperature. The present book is the end-product of two years' writing and research.

I begin with a history of thermometer development in Italy, France, Holland, Denmark, Sweden and England at the start of the modern age of scientific enquiry. After describing how temperatures have been measured over the land and the sea, and how reliant we still are on measurements made by simple old instruments, I move on to show how these measurements are processed to give average temperatures over small areas and how these are then converted into hemispheric and global averages. Throughout this enquiry, I avoid complex statistics in order to make the book intelligible to a wide audience, but with references for those who want more details of the maths.

Finally, I look into the pressing matter of what these measurements actually tell us about temperature change over the last few hundred years – coming to some unexpected and counter-intuitive conclusions. Using some of the best datasets currently available, I demonstrate that the situation regarding temperature change is complex and not reducible to one single graph or to one simple global truth. I show how we need to look at the local scale for clues as to how, and why, the climate is changing. Hemispheric and global values can be misleading because they merge many different climates and all four seasons, north and south, into one, thereby masking crucial details. I also explore doubts as to whether the merging of data from diverse climates and different times of the year has any validity in the real world.

In writing the book, I became increasingly aware that by stressing the complexity of the climate system, the uncertainty of the measurements and the intricacy of what they show, I run the risk of offending the committed hardline environmentalist and of adding fuel to the sceptics' case. I sit in the middle, unhappy with the often petulant tones of the sceptical debate while being unable to support the impression of certainty that many climatologists want to project, however worthy they may feel their motives are. But I end the book with positive suggestions as to how we might now proceed.

Acknowledgements

To avoid errors and the risk of being out of date, I sought advice and help from the following top experts in climate research:

Phil Jones (Climatic Research Unit, University of East Anglia) read four of my draft chapters and annotated them liberally, occasionally forcibly, for which I am most grateful; plain-speaking is what I wanted and is what I got. Many of his remarks, however, were exceedingly useful and I modified the text appropriately. On a few points we didn't agree, so any remaining mistakes or errors are mine. The book is very much the better for his contribution, so I hope we do not fall out over our disagreements.

David Parker (Hadley Centre, Met Office) also took the time to read some of the original drafts, including the chapter on sea surface and marine air temperature. David also gave valuable help with my questions regarding Central England Temperature and the diurnal temperature range, as well as on the latest research on warming in the troposphere. Like Phil Jones, I am sure David will not agree with everything I have written but would probably say that any remaining errors are mine.

Chris Folland (Hadley Centre, Met Office) has helped by supplying relevant recent papers and referring me to other researchers both in the UK and the USA where appropriate. Chris also suggested changing the subtitle of the book from my original *Past, Present and Future*, to the much better *Their Analysis and Interpretation*, which he believes conveys more accurately what the book is about – for the measurements are just the start, not the end. I think this is correct and I hope it encourages more people to read the book.

John Kennedy (Hadley Centre, Met Office) advised me on details of sea surface temperature measurements from ships and drifting buoys and on comparisons between them, supplying many relevant papers that ensured I was up to date. Climate science changes quickly and it is important to keep abreast of everyone's work.

Jon Turton (Hadley Centre, Met Office) read my draft section on the Argo float network, suggested changes and helped establish its temperature accuracy. Jon also advised on sea surface temperature and drifting buoys, augmenting what John Kennedy had provided, again ensuring I was up to date in what I said.

I am very indebted to all the above, who took the trouble to read what I had written. I appreciate the care they took and the time they spent doing this. Receiving their help made the task of writing the book more complex, since I had to rethink and rewrite several sections, but the book is that much the better for it. I have not agreed with all their comments, but that is no bad thing. I am sure I will be criticised for some of my comments and suggestions. Indeed I already have been.

I would also like to thank **Russell Vose** (National Oceanic and Atmospheric Administration, NOAA) for sending me the latest datasets on changes in the diurnal temperature range and for agreeing that I can use them in the book. These are invaluable data and I am very grateful to Russell for being so open and helpful. Indeed I have found that everyone I have approached has responded very positively, even when they disagreed with me.

I would like to thank my copy-editor, **Nancy Boston**, for the way she has picked up many small and not so small errors in what I had written. In effect, Nancy has been the first reader of the book and was able to draw my attention to points that I had not expressed clearly enough and mistakes I had overlooked. The book is very much the better for her care and attention, and the more readable for it.

I am also grateful to the staff of Cambridge University Press for publishing the book, especially to **Matt Lloyd**, who has been very supportive throughout the long process, from my initial proposal a few years ago, through the reviewing stage, followed by acceptance and, finally, publication.

Acronyms

AAO	Antarctic Oscillation
AMO	Atlantic Multidecadal Oscillation
AMSU	Advanced Microwave Sounder Unit
AO	Arctic Oscillation
AR4	Fourth Assessment Report (of the IPCC)
ATMS	Advanced Technology Microwave Sounder
ATSR	Along Track Scanning Radiometer
AVHRR	Advanced Very High Resolution Radiometer
AWS	automatic weather station
AXBT	airborne expendable bathythermograph
CAM	climate anomaly method
CDIAC	Carbon Dioxide Information Analysis Center
CEH	Centre for Ecology and Hydrology
CET	Central England Temperature
CLIVAR	Climate Variability and Predictability
CNES	French Space Agency
CRU	Climatic Research Unit
DBCP	Data Buoy Cooperation Panel
DCP	data collection platform
DTR	diurnal temperature range
ENSO	El Niño Southern Oscillation
ERI	engine-room intake
FD	first difference
FDM	first difference method
GCOS	Global Climate Observing System
GCR	galactic cosmic ray
GDP	Global Drifter Program

GEOSS	Global Earth Observation System of Systems
GHCN	Global Historical Climatology Network
GHG	greenhouse gas
GHRSST-PP	Global Ocean Data Assimilation Experiment High Resolution Sea Surface Temperature Pilot Project
GISS	Goddard Institute for Space Studies
GODAE	Global Ocean Data Assimilation Experiment
GOES	Geostationary Operational Environmental Satellite
GOOS	Global Ocean Observing System
GOSAMOR	Global Ocean Salinity Monitoring
GPS	Global Positioning System
GTS	Global Telecommunications System
ICOADS	International Comprehensive Ocean–Atmosphere Data Set
ICSU	International Council for Science Unions
IH	Institute of Hydrology
IOC	Intergovernmental Oceanographic Commission
IOD	Indian Ocean Dipole
IPCC	Intergovernmental Panel on Climate Change
IPO	Inter-decadal Pacific Oscillation
IPTS	International Practical Temperature Scale
IR	infrared
LAT	land air temperature
LEO	low Earth orbit
LIA	Little Ice Age
MAT	marine air temperature
MCDW	Monthly Climatic Data for the World
MHS	Microwave Humidity Sensor
MSS	multispectral scanner
MWP	Medieval Warm Period
NAM	North Annular Mode (or Arctic Oscillation)
NAO	North Atlantic Oscillation
NCAR	National Center for Atmospheric Research
NCDC	National Climatic Data Center
NMAT	night marine air temperature
NMS	National Meteorological Service
NOAA	National Oceanic and Atmospheric Administration
NOCS	National Oceanography Centre, Southampton
NODC	National Oceanographic Data Center
NORPEX	North Pacific Experiment

OWS	ocean weather ships
PDO	Pacific Decadal Oscillation
PMO	Port Meteorological Officer
PNG	Papua New Guinea
PRT	platinum resistance thermometer
QDO	quasi-decadal oscillation
RAM	random access memory
RCS	Reference Climatological Station
RH	relative humidity
RS	remote sensing
RSM	reference station method
RSS	remote-sensing systems
RTE	radiative transfer equation
SAM	Southern Annular Mode (or Antarctic Oscillation)
SAR	Second Assessment Report (of the IPCC)
SAR	synthetic aperture radar
SEVIRI	Spinning Enhanced Visible and Infrared Imager
SIO	Scripps Institution of Oceanography
SOC	Southampton Oceanography Centre
SOHO	Solar and Heliospheric Observatory
SOI	Southern Oscillation Index
SST	sea surface temperature
SVP	Surface Velocity Program
SVP	saturation vapour pressure
TAO	Tropical Atmosphere Ocean array
TAR	Third Assessment Report (of the IPCC)
TOGA	Tropical Ocean Global Atmosphere
TOPEX	Ocean Topography Experiment
UAH	University of Alabama in Huntsville
UHI	urban heat island
UNEP	United Nations Environment Programme
USHCN	US Historical Climate Network
UV	ultraviolet
VISSR	Visible and Infrared Spin-Scan Radiometer
VMO	voluntary marine observer
VOF	voluntary observing fleet
VOS	voluntary observing ships

VOSClim	VOS Climate Project
WCRP	World Climate Research Programme
WIS	WMO Information Service
WMO	World Meteorological Organization
WOCE	World Ocean Circulation Experiment
WV	water vapour
WWR	World Weather Records
XBT	expendable bathythermograph

1

The balance of energy

Solar radiation

The Sun emits around 74 million watts (W) of electromagnetic energy from each square metre (m^2) of its surface. The energy arriving at the outer boundary of the Earth's atmosphere, at right angles to the solar beam, is about 1,366 $W\,m^{-2}$ (watts per square metre). This value is known as the *solar constant*, although it is not quite constant, for reasons discussed later.

Figure 1.1 shows the spectrum of electromagnetic radiation emitted by the Sun, and also those parts absorbed during its passage through the Earth's atmosphere and the radiation received at the Earth's surface. Because the Sun is rarely directly overhead, and is out of sight at night, the average amount of energy received at any one point on the surface is much less than the solar constant; the global average at the outer boundary of the Earth's atmosphere being about 342 $W\,m^{-2}$ (or about one-quarter of the solar constant).

Passage of radiation through the atmosphere

During its passage to the ground (Figure 1.2), around 67 $W\,m^{-2}$ (or 19.6% of the 342 $W\,m^{-2}$ input) is absorbed by the atmosphere, warming it slightly in the process. A further 77 $W\,m^{-2}$ (22.5%) is reflected and scattered back to space by the atmosphere, clouds, aerosols, smoke from many sources, microscopic living organisms, dust from sandstorms, volcanic ash, micrometeorites from space, and salt crystals from the sea; the amount finally reaching the surface being 198 $W\,m^{-2}$ (58% of that received just outside the outer boundary of the atmosphere). A further 30 $W\,m^{-2}$ (9%) is immediately reflected back to space from the ground, the proportion reflected being known as the *albedo* of the surface, its magnitude depending on the nature of the surface. This means that the amount of energy lost back to space is about 107 $W\,m^{-2}$ (31.3% of the input) which, together with that absorbed

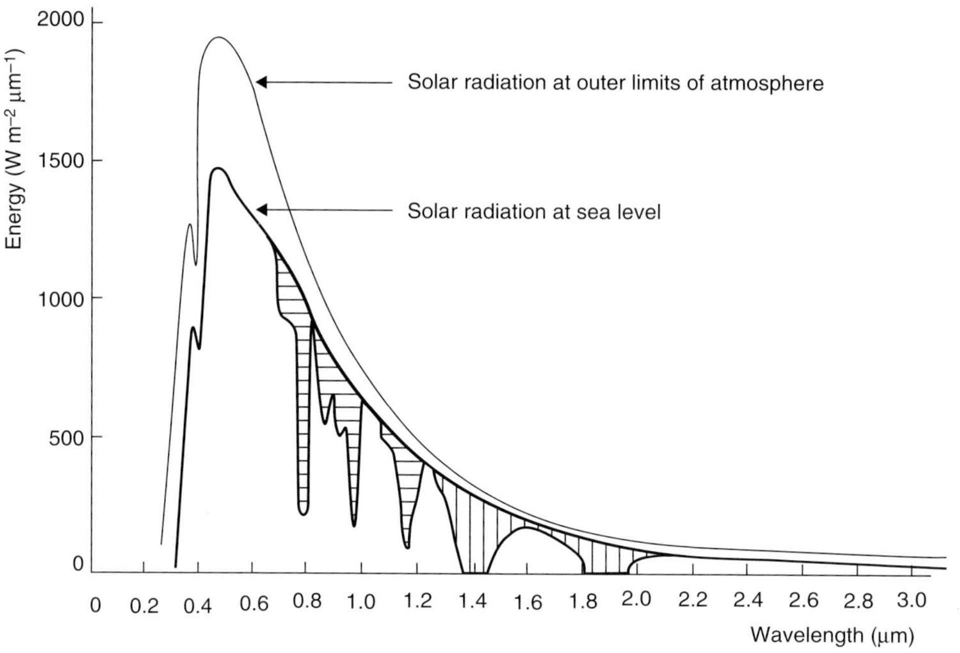

Figure 1.1 Spectrum of solar radiation at the outer limits of the atmosphere and on the ground (at sea level). *Vertical shading* indicates absorption by water vapour and carbon dioxide. *Horizontal shading* indicates absorption by water vapour and oxygen. Oxygen and ozone absorb in the UV below 0.4 µm (not shaded).

during its passage through the atmosphere, leaves 168 W m^{-2} 'available' at the surface (or 49.1% of the energy arriving at the outer boundary of the atmosphere). This is an average, as the levels available vary greatly with the time of day and season as well as with the nature of the surface. What happens to this 168 W m^{-2} depends on surface conditions – on whether it is land or sea, city or countryside, on the nature of the soil and of the vegetation, and on how much water is available.

Sensible heat flux

Initially the Sun's energy heats the Earth's surface, be it land or sea. Some of the warmth is then transferred to the air, first by molecular diffusion across what is known as the *laminar boundary layer* (a layer only a millimetre or so thick, which clings by molecular attraction to most surfaces). Beyond this, in the *turbulent boundary layer,* transfer of the warmed air away from the surface occurs through the action of the wind and eddies within it – a process that is much more effective than the initial very slow molecular diffusion. Through the action of wind turbulence alone, the atmosphere would be mixed rapidly to the top of the troposphere, taking the heat with it.

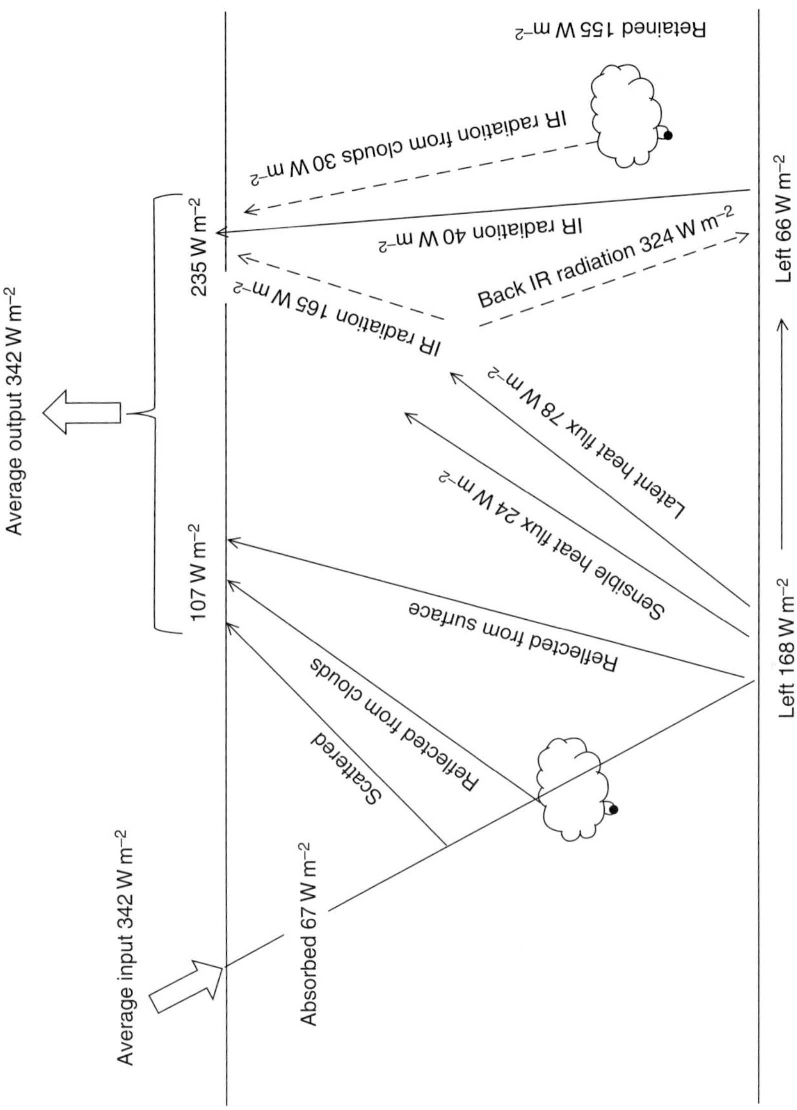

Figure 1.2 Energy exchange of solar and infrared (IR) radiation within the atmosphere. See text for discussion.

The atmosphere is made up of several layers (Figure 1.3): with the *troposphere* being the lowest layer, in which we live and in which most clouds exist, extending from the surface to a height of between 8 and 20 km, depending on latitude; its upper limit is known as the *tropopause*. The transfer of warmed air away from the surface is known as the *sensible heat flux* ('sensible' in that it can be sensed, or measured, as a change in temperature; 'flux' referring to the rate of passage of the heat energy upwards). The sensible heat flux consumes, on average, 24 W m^{-2} (14.3%) of the available 168 W m^{-2} solar input, but this varies greatly with time, place and conditions.

Soil heat flux

Over the land, a small proportion of the heat from the warmed surface is also transferred downwards by conduction as the *soil heat flux*, this generally being returned to the surface during the night or over the course of a season. The rate of transfer of the warmth downwards is influenced by the amount of water in the soil, by the soil's pore and particle sizes, and by the presence of vegetation. Many biological processes are influenced by soil temperature, from the activity of micro-organisms to the germination of seeds and plant growth. Over the oceans somewhat different processes occur (see the section *Radiation processes over the oceans* later in this chapter).

Latent heat flux

In addition to warming the atmosphere, a proportion of the available solar energy arriving at the Earth's surface also evaporates some, or all, of the available water, producing water vapour. The resultant vapour mixes with the warmed air and is transported away by the same processes. This is known as the *latent heat flux* ('latent' because the heat is hidden – there being no change of temperature associated with the process, just a change of state from liquid to gas). For reasons explained later, this hidden energy is important. Evaporation takes place from the sea, from open lakes and rivers, from moist soil, from wet vegetation and as transpiration from plants through their stomata; the whole process often being called *evapotranspiration*. An average of 78 W m^{-2} is used up in this way, although the amount depends very much on circumstances, in particular on the amount of water actually available, on the temperature of the air and water, on how humid the air already is, on wind speed, on the texture and colour of the plants and soil, and on the many different combinations of these factors. Over the sea, for example, the unlimited supply of water produces a somewhat different relationship between sensible and latent heat flux than that found above a desert.

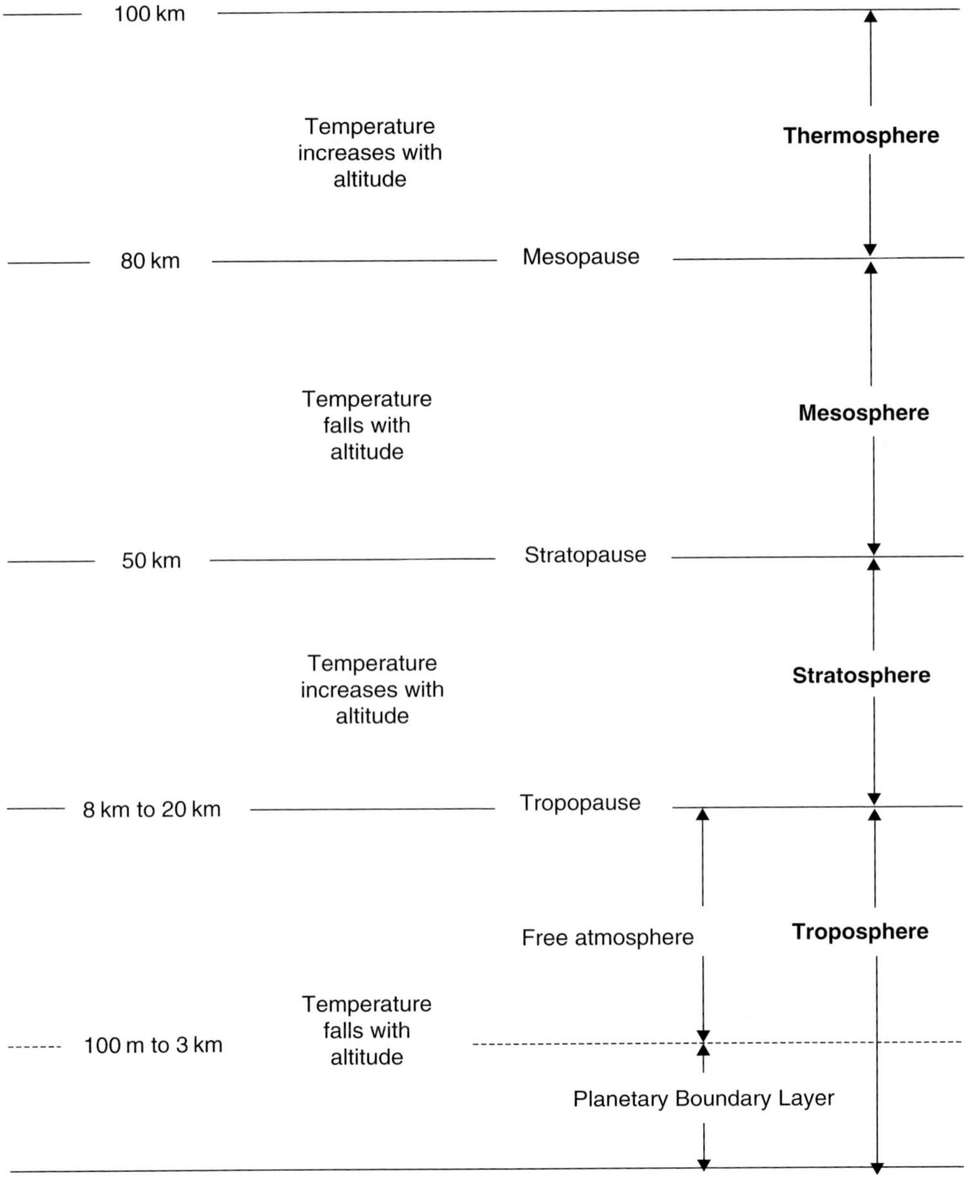

Figure 1.3 Structure of the atmosphere.

Convection and clouds

As well as being transferred away from the ground by wind turbulence, the warmed air is also transferred upwards by the process of *convection*: warmer, and therefore less dense air rising into the relatively cooler, denser air above, taking the water vapour with it. To balance out this upward movement, cooler air descends elsewhere, thereby thoroughly mixing the atmosphere. This mixing occurs right up to the tropopause (Figure 1.3). (There is also a process termed *advection*, which is simply the movement of a fluid in any direction, such as sideways as in the blowing wind; it means 'to transport away'.)

The next piece of reasoning is important. Because air pressure reduces with height, rising packets of air (often known as *thermals*) expand as they ascend and, in so doing, they cool by virtue of the first law of thermodynamics – the conservation of energy. (Those interested in a mathematical explanation of why this happens can refer to Boyle's law and Charles's law in Appendix A.) The process cannot be understood intuitively, but it is not necessary to be able to follow the (simple) maths in order to accept the fact that the expansion of a gas cools it, and vice versa (for instance, a bicycle pump heats up when inflating a tyre, due to the air inside being compressed). Until the maths was done, the whole process was something of a mystery to scientists (Strangeways 2007). Provided that the rising thermals do not mix (much) with the air that they are rising through, the rate of cooling is known as the *dry adiabatic lapse rate* and is between 6° and 10°C for each kilometre increase in height. When mixing of the air occurs, it is termed *diabatic* or, more usually, the negative term *non-adiabatic*. 'Lapse' refers to the fact that the temperature decreases with height.

Because of this cooling, a height will eventually be reached where the temperature of the rising thermal becomes the same as the surrounding air, whereupon ascent stops. However, the water vapour which has risen with the warmed air may condense back into water (as cloud droplets) if the air cools to the *dew point* (see Appendix B). The important point here is that when the water vapour condenses back to the liquid form, all of its 'hidden' (latent) energy, absorbed in the evaporation process at ground level, is now released, thereby warming the air. Thus heat is transported upwards in the guise of water vapour as well as in the form of warmed air, provided, that is, that condensation occurs. Moreover, once condensation starts to occur, the resultant warming stops the air from cooling as fast as it would have done if there had been no condensation, and this has the effect of causing the thermal to continue rising further than it would have done if the dew point had not been reached. The lapse rate of moist air, known as the *wet adiabatic lapse rate*, is thus less than that of dry air (about 4.5–5.0°C per kilometre rise) because the heat released by condensation partly reduces the rate of cooling. Under certain atmospheric conditions, termed 'unstable', the thermal can continue to rise right to the top of the troposphere, where it meets the tropopause.

The tropopause

The tropopause is the boundary between the top of the troposphere and the bottom of the stratosphere (Figure 1.3), at which point rising thermals stop ascending because they meet a 'ceiling' or inversion. The ceiling results from the fact that, in the stratosphere, temperature increases with height – in contrast with it decreasing with height in the troposphere – thereby reducing or preventing convection. At the tropopause, therefore, the air and any clouds within it spread out sideways, travelling downwind, producing the familiar anvil shape of a thundercloud (Strangeways 2007).

The rise of temperature with altitude within the stratosphere is due to heating from above by the absorption of ultraviolet (UV) wavelengths by the ozone layer – this action giving off heat. The ozone itself is formed in the first place by the UV radiation splitting oxygen molecules (O_2) into single atoms of oxygen, these then combining with O_2 molecules to form O_3, or ozone. The UV then subsequently splits the O_3 back into O_2 and O, with the process cycling continuously. Most of the ozone is concentrated at an altitude of between 15 and 35 km, where its concentration is between 2 and 8 ppm (parts per million), although the ozone extends from 10 km to 50 km. The amounts are very small; if all the ozone was collected together at surface pressure it would be only a few millimetres thick.

The tropopause occurs where the heating by conduction from above, in the ozone layer, and the heating by convection from below balance out. Because the surface is colder in polar regions, the tropopause is lower there. But even within the troposphere, convection may be halted by inversions before the tropopause is reached, because under some conditions, such as an anticyclone, warm air can occur above colder air below, producing what is termed a 'stable' atmosphere.

As noted earlier, the latent heat transfers just under 78 W m^{-2} of the available 168 W m^{-2} arriving at the Earth's surface, leaving, on average, 66 W m^{-2} still unaccounted for. What happens to this remaining energy?

Infrared radiation and the greenhouse effect

Until Sir Benjamin Thompson, Count Rumford, wrote his classic paper (Thompson 1804) on infrared (IR) radiation, no-one was aware of its existence, nor of the important role that it played in meteorological processes. In particular, no-one could explain why dew formed on clear nights. What Count Rumford found, by experiment, was that all objects above a temperature of absolute zero (see the section *The absolute, or thermodynamic, temperature scale* in Chapter 2) emit IR radiation and, in so doing, are cooled. Infrared radiation is that part of the electromagnetic spectrum (see Appendix C) which has a wavelength longer than that of visible light but shorter than that of microwaves. Everything above a temperature of absolute zero emits IR. (When an object is very hot, some of the radiation emitted is in the visible spectrum, the hotter the bluer, the cooler the redder.) At lower temperatures, the radiation is restricted to the IR and microwave spectra; the actual range of wavelengths and their intensity depends on the temperature of the object. The IR spectrum stretches from 0.7 μm to 1 mm, divided into near-, mid- and far-IR (Appendix C).

The remaining 66 W m^{-2} of the available energy at the surface is radiated from the warmed ground in the form of this IR radiation, with the amplitude and wavelength depending on the temperature of the ground. Most gases are not affected by the passage of IR radiation through them and, if the atmosphere was composed entirely of oxygen and nitrogen, the IR radiation would pass through as if they were not there and would be lost to space; there would be no greenhouse effect. But some gases do react with IR by absorbing the radiation. In Earth's atmosphere the main greenhouse gases (GHGs) are water vapour (WV), carbon dioxide (CO_2) and methane, with WV being by far the most powerful and prevalent GHG, accounting for about 95% of the total greenhouse effect, carbon dioxide accounting for 3.6% and methane and all the others put together accounting for 1.4%.

The GHGs in the atmosphere absorb some of the IR emitted from the surface, and warm up. But it is in the nature of molecules that absorb radiation at a given frequency also to radiate it at the same frequency, so the warmed GHGs re-radiate some of the received heat in all directions, part of it downwards towards the surface, causing the atmosphere below, and the surface, to warm up. Some of the upwardly re-radiated IR is lost to space and some is absorbed by the atmosphere higher up; it is a very circular process at all altitudes. The amount of IR radiation received back at the surface, as *back radiation*, from all of the GHGs mentioned above, is about 324 W m^{-2}, which, when added to the 66 W m^{-2} remaining from the incoming energy budget (see above), amounts to 390 W m^{-2}. Continuing the circular process, this radiation is now radiated back upwards.

Because the planet must be in thermal equilibrium, the energy lost to space must balance that coming in from the Sun. Since the input from the Sun is 342 W m^{-2}, of which 107 W m^{-2} is lost as reflected solar radiation, the infrared losses must make up the difference; that is $342 - 107 = 235$ W m^{-2}. This is made up of three parts: a direct loss of 40 W m^{-2} through a clear atmospheric 'window' (a band of wavelengths transparent in that part of the IR spectrum), a loss of IR from cloud tops of 30 W m^{-2}, and a loss of 165 W m^{-2} from the atmosphere, totalling the required balance of 235 W m^{-2}. This means that the energy retained by the greenhouse effect is $390 - 235 = 155$ W m^{-2}, keeping the surface and lower troposphere warmer than it would otherwise be by about 21°C. Without the GHGs, the average temperature of the planet would be −6°C. Thus, with the GHGs, the average is 15°C.

Since atmospheric temperature, overall, decreases with altitude, the radiation emitted becomes less intense with increasing height. Thermal balance is, therefore, achieved by the temperature of the Earth's surface and its lower atmosphere increasing sufficiently for the cold upper atmosphere on the edge of space to be warm enough to radiate exactly the required 235 $W m^{-2}$. So it is the drop of temperature with altitude that makes the greenhouse process work. It would not happen if the temperature was the same all the way up. Nor would it happen if the atmosphere was composed only of oxygen and nitrogen, since they do not absorb IR radiation. Indeed, if this were so, life would probably not have developed on Earth, for the surface would be at an average temperature of −6°C. We should, therefore, be thankful for the presence of water vapour and carbon dioxide; our existence depends on them. (While the solar input only occurs during the day, the IR radiation is continuous, day and night.)

The greenhouse effect has been going on ever since the Earth's atmosphere formed 4.5 billion years ago, when it contained water vapour, hydrogen, hydrogen chloride, carbon dioxide, carbon monoxide and nitrogen. Through its interaction with rocks and early living organisms, the primitive atmosphere slowly evolved, with photosynthesising plants eventually changing some of the carbon dioxide into oxygen. The atmosphere reached its present composition only about 280 million years ago, since when it has changed but little. Its fixed part is now made up of 78.09% nitrogen, 20.95% oxygen, 0.93% argon and 0.03% carbon dioxide (by volume), along with a few trace gases.

The water vapour component of the atmosphere, however, varies considerably across time and space. The amount of water vapour that air can hold (the *relative humidity*: see Appendix B) reaches a theoretical maximum of about 3%, although normally it is less, with 2% representing very moist air. The presence of water vapour in the atmosphere causes complications when considering the greenhouse effect, because it is a considerably more powerful GHG than carbon dioxide and a great deal more plentiful, consequently producing a much greater greenhouse effect than carbon dioxide (about 95% of the total).

The *enhanced greenhouse effect* is the increase in retained heat above the natural mean of 155 $W m^{-2}$ (see above) as carbon dioxide levels increase, amounting to about 4 $W m^{-2}$ for a doubling of the pre-industrial concentration. Although there is no doubt whatsoever about the physics and maths of the greenhouse effect, CO_2 does not act on its own and there are many other factors involved in the workings of the climate. Consequently a very simple, one-to-one, direct link between the amount of CO_2 present in the atmosphere and the temperature at the Earth's surface is most unlikely to exist. It is, however, most probable (all else being equal) that an increase in temperature will increase evaporation and thus the amount of water vapour in the atmosphere. Being a GHG, the water vapour might then increase the temperature further. In climate models water vapour is treated simply as a positive feedback in isolation, resulting from temperature increases due to CO_2 increases. However, the increased water vapour might increase low cloud cover and this would reflect more solar radiation resulting in negative feedback and cooling. Some models include this effect, others do not (see *Clouds*, page 149). The intertwined processes producing Earth's climate are complex.

Measuring the radiation

Sunshine duration

It was not possible to measure the output of the Sun precisely until there were satellites above the atmosphere equipped with suitable sensors. Prior to this, measuring the intensity of solar radiation at the Earth's surface was all we could do. Until the last 50 or so years, the most common way of measuring solar activity was by the Campbell–Stokes sunshine recorder developed in the nineteenth century (Strangeways 1998, 2003) (Figure 1.4). Although elegant and well designed, it gives only an indication of whether the Sun is shining brightly or is obscured by clouds. It tells us only indirectly about the radiation intensity: only if the time of day, date and latitude are known can a fair estimate of intensity be calculated. It is also difficult to interpret the burnt tracks on the recorder charts, especially if clouds obscure the Sun frequently for short periods of time. The recorder does, however, give a useful record of cloudiness, but only during the daytime. Modern electrical equivalents of this old device have been developed to enable sunshine duration to continue to be measured by unattended automatic weather stations (AWSs).

Solar radiation

To measure the intensity of solar radiation, instruments known as *pyrheliometers* automatically track the passage of the Sun across the sky, in the same way as a telescope on an equatorial mount tracks stars. The second-by-second variations in intensity of the radiation are measured by sensing the temperature of a black surface exposed to the Sun at the bottom of a tube, relative to a similar black surface screened from the Sun (thereby compensating for air temperature changes). The tube has a field of view that encompasses just the Sun's disc and a small annulus of sky around it. The variations in intensity caused by changes in the atmosphere would swamp any changes occurring in the amount of energy

Figure 1.4 Campbell–Stokes sunshine recorder. The trace burnt on the card gives a rough indication of sunshine duration.

arriving from the Sun outside the atmosphere and so it is not possible to deduce, with any certainty, by how much, if anything, the Sun's output is changing.

Total incoming solar radiation

In addition to pyrheliometers, which measure just the direct beam at right angles to the Sun, instruments known as *pyranometers* (Figure 1.5) measure the total incoming solar radiation on a horizontal disc protected by two concentric glass domes. Two domes are used to reduce the effect of IR radiation from the outer dome being added to the incoming solar radiation. The disc is black to absorb the radiation across the full spectrum, with the radiation warming the disc by an amount that is dependent on the radiation intensity. The temperature of the disc is measured relative to the air temperature beneath the white surrounding screen by means of a thermocouple. The incoming radiation includes the direct solar beam and all the diffuse solar radiation from the whole dome of the sky, including that from a

Figure 1.5 Pyranometers measure the total incoming solar radiation, both directly from the Sun and that scattered and reflected from the atmosphere and clouds.

cloudless blue sky and that reflected off clouds and scattered by haze. This total can actually be greater than the solar constant for a time, if clouds act as reflectors. Overall, of course, clouds reduce the amount of radiation received at any point on the ground.

Net radiation

It is not just solar radiation that arrives at the surface. As shown earlier, infrared radiation arrives back at the surface from the warm atmosphere and clouds above. The total incoming energy across the full spectrum minus the total outgoing solar and infrared radiation is known as the *net radiation*. This is an important variable because it is this combined total radiation that is available to warm the ground and air and to evaporate water. In measuring the energy balance at the Earth's surface, therefore, the measurement of net radiation is more important than simply measuring the incoming solar radiation.

Net radiation is measured by *net pyrradiometers* (Figure 1.6). Whereas pyranometers sense only radiation of solar origin in the fairly narrow spectral band from 0.3 to 3.0 μm, and only look upwards, net pyrradiometers measure the much wider range of about 0.3–70 μm from both above and below. This requires the use of polyethylene domes to protect the sensing surfaces in place of the glass domes of pyranometers, as glass is not transparent to the longer wavelengths. The sensor must also be double-sided, having both an upward- and downward-looking black sensing surface, in order to measure the difference between all incoming and all outgoing radiation. Whereas glass is a stable, easily cleaned and tough material, the thin polyethylene domes that

Figure 1.6 Net pyrradiometers measure the difference between the total incoming and total outgoing solar and infrared radiation; that is, the total exchange of electromagnetic radiation.

have to be used with net sensors are very vulnerable to aging through the action of UV, and to weathering generally; they are also more difficult to keep clean. Fuller details of all the radiation sensors can be found in Strangeways (2003).

Radiation processes over the oceans

The same radiation processes as those described earlier over the land also occur over the oceans. However, due to the availability of unlimited water, a different balance of the three kinds of fluxes will result, depending on the temperature of the sea surface and of the air, on the relative humidity of the atmosphere, on the sea surface conditions and on wind speed. The action of soil heat flux is now replaced by the flux of energy that warms the sea, and this behaves differently from the ground since the sea is semi-transparent, fluid, and moves both laterally and vertically, with waves mixing the surface waters. The visible radiation enters the water and penetrates downwards for several tens of metres, being absorbed in the process – warming the surface waters directly rather than simply by conduction, as over land. Despite this complication, the overall situation is simpler over the sea because there is an unlimited supply of water at all times and because there are no hills, no deserts, no cities and no vegetation cover. The main differences between radiation processes over land and over the oceans can be explained in terms of the proportions into which the sensible and latent heat fluxes are divided.

Changes in solar activity and climate change

Whether changes in solar activity affect the climate is a controversial question, in part because there are not yet enough hard facts to go on. What follows is a brief review of the essentials of the argument.

Milankovitch cycles

There are long-term changes in Earth's orbit that affect the amount of radiation received, notably the cycles described by Milankovitch (1941). These do not involve changes in the level of solar radiation emitted by the Sun, but simply the amount of radiation received by Earth due to changes in its angle of rotation and distance from the Sun. These changes most probably account for the arrival and departure of the Ice Ages. On a shorter time scale, they will have no effect on climate.

Sunspots

The existence of sunspots was known to the Chinese in 800 BC, and the first known drawings date from the twelfth century. Precise records have now been kept for several hundred years since the invention of the telescope by Galileo. Heinrich Schwabe, working between 1826 and 1843, was the first to detect an 11-year cycle in sunspot numbers (Schwabe 1844), and this led Rudolf Wolf in 1848 to develop a method of quantifying the level of solar activity, the *Wolf number* (or *Zürich number*) becoming adopted as an expression of the number of individual spots and groups of spots (Lamb 1972). Gustav Spörer noted that there were very few sunspots in the fifteenth and sixteenth centuries – this phenomenon was given the name the *Spörer minimum* (Lamb 1972). Edward Maunder then found that from 1645 to 1715 there were very few sunspots, with virtually none from 1672 to 1699, this becoming known as the *Maunder minimum* (Lamb 1972). During this minimum, the solar corona was smaller and aurora displays were few. Maunder also observed that the low sunspot count coincided with low temperatures. The Little Ice Age (LIA) started somewhere around 1350 and ended around 1850, with particularly cold periods from 1425 to 1550 and from 1650 to 1715, coinciding with the Spörer and Maunder minima. The Medieval Warm Period (MWP) (1100–1300) occurred during a period of high solar activity. While association is no proof of causation, these coincidences suggest that there is perhaps some connection between sunspot number (indicating solar activity) and climate, although the mechanisms are still uncertain.

William Herschel (1801) proposed that sunspot activity could be related to the price of wheat, because it affected the mildness or severity of terrestrial seasons and hence the success or failure of global harvests.

If changes in temperature and number of sunspots *are* linked, through simple (or complex) cause and effect, and since the Sun's activity is global in influence, it follows that the LIA was potentially also global, although perhaps modified by local conditions and so not necessarily expressed everywhere worldwide, or perhaps expressed differently. These periods in history are well documented in Europe but less so elsewhere, especially in the Southern Hemisphere. What solar processes might have caused these swings in temperature?

Changes in the solar constant

The solar constant is not constant. It changes in two different ways.

Changes due to Earth's elliptical orbit

Although the solar constant is quoted as being 1,366 W m^{-2}, this is an average over the year. Because Earth is in a slightly elliptical orbit round the Sun, rather than a circular one, the distance to the Sun varies over 12 months. We are closest to the Sun on 3 January (distance 147,166,462 km) and furthest away on 4 July (152,171,522 km), a change of 5,005,060 km or 3.3%. Because of the inverse square law, the change in radiation received by Earth over the year, resulting from its change in distance to the Sun, is 85.78 W m^{-2} or 6.3% of the mean solar constant. Thus the Northern Hemisphere receives more radiation in its winter months than the Southern Hemisphere does in its winters, and, for the same reason, the Northern Hemisphere receives less radiation in its summers than the Southern Hemisphere does in its summer months. To put this change into context, the natural greenhouse effect has a mean value of 155 W m^{-2} while the enhanced greenhouse effect amounts to about 4 W m^{-2} for a doubling of the pre-industrial concentration (see *Infrared radiation and the greenhouse effect* earlier). The seasonal difference in received solar energy between the two hemispheres is thus very large in comparison, and this will be reflected in different climates North and South. The significance of this when merging the two hemispheres to produce a global mean temperature, and in attempting to establish the causes of climate change, is discussed in the section *Are anomalies weightless* in Chapter 8.

Changes in the Sun itself

Before the advent of satellites, the solar constant was known only roughly from measurements made on the ground (see *Measuring the radiation* above). Over the last 30 or so years, measurements have been made from satellites and the solar constant has been found to change by only about ±0.05% from 1366 W m^{-2}, peaking when sunspot numbers are at a maximum. This is a measurement across the full solar spectrum, averaged over the year (see previous paragraph). Changes of this small magnitude probably cannot affect the climate. What other solar characteristics might have caused the apparent climate-dependence on sunspot numbers?

Changes in the Sun's ultraviolet output

While the overall level of solar radiation is very stable, the SOHO (Solar and Heliospheric Observatory) satellite showed that, during an 11-year solar cycle, changes in the ultraviolet (UV) part of the spectrum are proportionally much greater than changes across the full spectrum, with ranges varying widely through factors of 2–10 times the minimum during a solar cycle. UV reacts with oxygen to produce ozone (see *The Tropopause* section above), heating the stratosphere in the process. An increase in UV increases stratospheric heating, changing the temperature structure not only of the stratosphere but also of the circulation in the troposphere beneath (IPCC 2007). This can influence the Arctic Oscillation (AO), forcing depressions further south, causing lower latitudes to cool, or to move further north causing them to warm, thereby affecting the temperatures observed at the Earth's surface. Its effects on the Antarctic Oscillation (AAO) are less well known. Nearly 40 years ago, Hubert Lamb, one of the principal pioneering climate scientists of the time, wrote extensively on the topic of variations in UV radiation from the Sun and its possible effects on the climate (Lamb 1972), examining the complex processes that might be involved. Yet the matter has still not been resolved satisfactorily.

The Sun's magnetic field and cosmic rays

Another way in which the Sun could affect temperatures is through its magnetic field and the *solar wind* (Pallé et al. 2004; Tinsley and Yu 2004). The Sun's atmosphere is a plasma continually radiating protons and electrons into space, these particles travelling to the outer edge of the solar system. Since the solar wind is also a plasma, it is electrically conductive, and this causes the Sun's magnetic lines of force to be carried along with it. This creates a 'shield' or 'bubble' around the Sun in the interstellar medium (the hydrogen and helium gas between the stars). At the *heliopause*, well beyond the orbit of Pluto, the shield is no longer powerful enough to hold back the interstellar particles.

The magnetic field and the solar wind deflect some of the incoming galactic cosmic rays (GCRs), with more being deflected during a sunspot maximum. GCRs are the primary source of ionisation in the troposphere (above 1 km), and changes in GCR numbers *may* affect the number of nuclei available for cloud formation (Strangeways

2007); a weak sun allows more GCRs to reach Earth, possibly causing more clouds to form, reducing the input of solar energy and thereby lowering the temperature. There is, as yet, no direct experimental proof of the influence of GCRs on cloud formation, but experiments are planned, using a particle accelerator to simulate the effect.

There is a convenient by-product of GCRs. As they enter the atmosphere, the protons amongst them can combine with nitrogen atoms to produce the heavier form of radioactive carbon-14, or they can break heavy atoms apart to produce the lighter beryllium-10. These isotopes find their way into tree rings and ice cores and, by measuring their numbers at different times in the past, it is possible to trace back a long history of cosmic ray intensity and, therefore, of solar activity and sunspot numbers. Such measurements have confirmed the 400 years of visual observations and have also extended the record of sunspot numbers back still further.

In conclusion, it appears that variations in the Sun's activity could possibly influence global temperatures in at least two ways – and possibly in ways we have not yet thought of. The IPCC (2007) have concluded that 'More research to investigate the effects of solar behaviour on climate is needed before the magnitude of solar effects on climate can be stated with certainty'. This is still very much a 'work in progress'.

References

Herschel, W. (1801) Observations tending to investigate the nature of the Sun. *Philosophical Transactions of the Royal Society of London* **91**, 265–318.

IPCC (2007) *Climate Change 2007: The Physical Science Basis.* Contribution of Working Group I to the Fourth Assessment Report of the Intergovernmental Panel on Climate Change. Cambridge University Press.

Lamb, H. H. (1972) *Climate: Present, Past and Future.* Methuen: London.

Milankovitch, M. (1941) *Kanon der Erdbestrahlungen und seine Anwendung auf des Eiszeitenproblem.* English translation by Pantic, N. (1998) *Canon of Insolation and the Ice Age Problem.* Alven Global: Belgrade.

Pallé, E., Butler C. J. and O'Brien K. (2004) The possible connection between ionization in the atmosphere by cosmic rays and low level clouds. *Journal of Atmospheric and Solar-Terrestrial Physics* **66**(18), 1779–90.

Schwabe, S. H. (1844) Sonnen-Beobachtungen im Jahre 1843. *Astronomische Nachrichten* **21**, 233.

Strangeways, I. C. (1998) Back to basics: The 'met enclosure'. Part 3. Radiation. *Weather* **53**(2), 43–9.

Strangeways, I. C. (2003) *Measuring the Natural Environment* (2nd edn). Cambridge University Press.

Strangeways, I. C. (2007) *Precipitation: Theory, Measurement and Distribution.* Cambridge University Press.

Thompson, B. (1804) An enquiry concerning the nature of heat and the mode of its communication. *Philosophical Transactions of the Royal Society of London* **94**, 77–182.

Tinsley, B. A. and Yu, F. (2004) Atmospheric ionization and clouds as links between solar activity and climate. In: *Solar Variability and its Effects on Climate*, Geophysical Monograph **141**. American Geophysical Union: Washington DC, pp. 321–39.

2

Thermometry

This chapter explores the development of thermometers from the time of the first experiments up to present-day instruments. The next two chapters then look at how thermometers have been used over the past 150 years to measure land air temperature. Chapter 5 explores similar measurements over the oceans. Thereafter it is a matter of addressing how these land and sea measurements can be converted into global temperature trends and what these trends indicate. With most scientific inventions, there is often uncertainty as to who did what first, and thermometers are no exception.

Air thermometers

Philo of Byzantium and Hero of Alexandria

As with many such matters, the earliest signs of human thought about temperature were first recorded in the Ancient World, with the work of Philo of Byzantium (late second century BC) and Hero of Alexandria (first century AD). These men designed many mechanical devices and, although their writings were lost for centuries, throughout the 'Dark Ages' and medieval era in Europe, some of them had been conserved by the Arabs during that time. Philo's notes re-emerged in the nineteenth century. Hero's notes had reappeared a little earlier, in Italy in 1575, first in Latin and then, in 1592 – the year in which Galileo started work at Padua – in Italian.

In deciding who made the first thermometer, it is important to differentiate between a thermoscope and a thermometer; the latter being the former with a scale added to it – a 'meter' must have a scale. Then it needs to be asked when a pneumatic experiment becomes a thermoscope – that is, something designed specifically to give a visible indication of changes in temperature, rather than something that does so as an incidental by-product. Such pneumatic experiments were performed

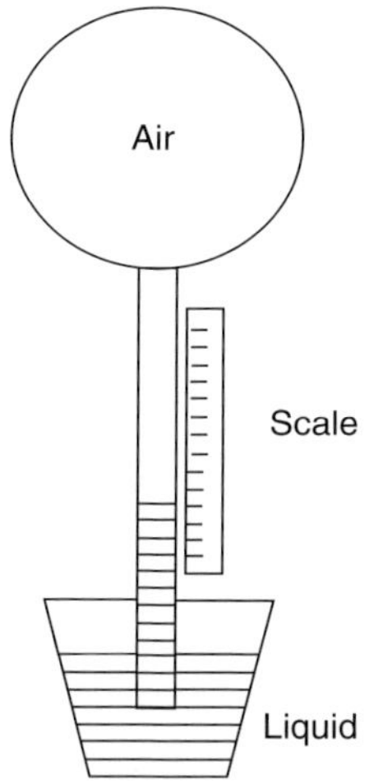

Figure 2.1 An air thermometer measures the change in volume of air in the bulb as the temperature changes, the liquid moving up and down as the temperature fluctuates. Without a scale, it is a 'thermoscope', with a scale it is a 'thermometer'. However, the reading is also affected by barometric pressure.

by both Philo and Hero, but there is no sign that either of the men used them to give even an indication of temperature, let alone a measure of it. Hero's 'fountain that drips in the sun' was not a thermoscope but something to show that 'water could be raised by heat'.

Della Porta and Santorio in Italy

Around 1,600 years later, these ancient devices and curiosities were pondered over in Italy by Galileo and della Porta, using the newly come-to-light literature, the latter copying Hero's design – but only to show the expansion of air by heat. Then around 1612, Santorio Santorio (also known as Santorio Santorii, or Sanctorius of Padua, 1561–1636) applied a measuring scale to an air thermoscope, making an air thermometer. There is some uncertainty as to whether Galileo also made such a device and, while arguments abound, with evidence in various writings, there is no certainty either way; but it is not, in any case, particularly important. The general principle of an air thermometer is shown in Figure 2.1.

Fludd in England and Drebbel in Holland

In a search for the chronology of the emergence of the first thermometer, we must also look at the work of the Welshman, Robert Fludd (or Flud), a doctor working in London, and at the activities of the Dutch inventor Drebbel. In Fludd's book

Meteorologica Cosmica (1626) a drawing clearly shows what can be described as an air thermometer (as shown in Figure 2.1) or a 'weather glass', as it soon became known. He seems to imply in his writings that he invented it, although Francis Bacon had given full details of how to construct such an instrument six years earlier. By 1634 it was in commercial production and was a familiar object in England by 1638.

Air thermometers were also the province of Otto von Guericke of Magdeburg in Germany and Sir Christopher Wren, and by the mid seventeenth century they were widespread. But who made the first air thermometer is still not clear, although Santorio is by far the most likely candidate. It could be that it was invented several times independently, since the mood of the time was experimental and dynamic in Europe: all of this took place between about 1612 and 1630. Notwithstanding who was first, the idea spread rapidly.

However, air thermometers suffer from a serious defect: the height of the column of water also responds to changes in barometric pressure. This was realised soon after Torricelli's work on mercury barometers around 1644 (Middleton 1994). Indeed, a very similar construction had been around for 50 years before the Torricelli experiment, and before thermometers were made, known as the *Dutch weather glass*, which was in effect an air thermoscope (Strangeways 2003) although it was not called that, nor called a barometer, because the groundwork had not yet been done to give it a name. So, in a sense, the air thermoscope and the barometer had been in existence in one instrument for half a century before it was called either, its significance being vague and its use mainly decorative. In his book, *A History of the Thermometer and its Uses in Meteorology*, Middleton (1966) does not mention this device, perhaps because he did not discuss it in his earlier book *The History of the Barometer* (Middleton 1964) either, maybe looking upon it as a toy; but its existence does show how the order of events is not always clear-cut. The Dutch weather glass did change with the weather, but no-one knew why.

Liquid-in-glass thermometers

Jean Rey

The next development came from Jean Rey in the Dordogne in France around 1630 who, totally unaware of the air thermometer and without any outside help, made the first ever liquid-in-glass thermometer. This was a bulb containing water with a narrow-bore tube above, into which the water expanded as it warmed (Figure 2.2). But it was open at the top and therefore not a permanent device. Nevertheless, the idea had been generated, by one person working alone. Jean Rey can be credited with inventing the liquid-in-glass thermometer.

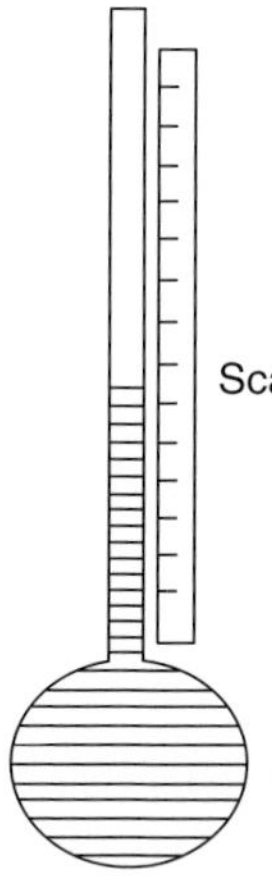

Figure 2.2 A liquid-in-glass thermometer measures the change in volume of the liquid, the liquid moving up and down the capillary bore of the tube as the temperature fluctuates.

Ferdinand II, Grand Duke of Tuscany

The first known sealed liquid-in-glass thermometer was made, without any doubt it seems, in Florence by the Grand Duke of Tuscany, Ferdinand II, one of the Medici family.

Two kinds of thermometer were invented in Florence. One consisted of a tube filled with spirits of wine (multiple-distilled ethyl alcohol, also known as ethanol) and containing a number of glass balls each with a different weight-to-volume ratio (density). This was first made in or before 1641. At low temperatures, when the density of the ethanol is high, all the balls float. As the temperature rises, the balls sink one after another. Similar devices are now sold in gift shops under the (presumably incorrect) name of *Galileo thermometers*. The second type of thermometer invented in Florence by Ferdinand II was the type familiar to us today, in which ethanol fills a bulb and expands into a narrow-bore tube as the temperature rises. Ferdinand's, however, was sealed at the top and so represents the prototype of today's instruments. Documents studied by Middleton (1966) in Italy date this instrument as originating around 1654 or a year or two earlier. The thermometer is closely associated with the *Academy of Experiments* (Accademia del Cimento) set up by the Grand Duke in 1657. In its short (10-year) existence, the Academy could be said to have paved the way for modern experimental physics, but it was closed in 1667, probably through opposition from the Church, which felt uneasy about all scientific enquiry, as we know from Galileo's treatment by the Church over his astronomical discoveries made with the first precision telescope. The Academy's demise ended Italy's predominance in the emerging new sciences of the seventeenth century. While the liquid-in-glass thermometer had first been made before the Academy opened, it was the first object to be studied experimentally there.

The original thermometer had a scale of 100 steps, but there was a smaller model of 50 and a larger one of 300 steps; several of the smaller models still survive in Florence, a chest-full, formerly belonging to the Academy, having been discovered in 1828. These scales were not based on any fixed reproducible temperature points, although the amount of liquid in the tube was set as a last step of construction, using ice and the Sun as opposing points, but only as rough temperature guides; there was no appreciation that the ice-point was of fundamental significance, repeatable and stable: this came much later.

Ethanol was used as the liquid because it changed more in volume with temperature than water does. Water was also thought to 'spoil the transparency of the tube' and it was also found that it reached its minimum volume slightly above freezing point (not to mention the inconvenience of it actually freezing). The spirit was initially coloured red but this was later stopped when it was found that the dye gradually coloured the glass beyond the liquid, making it hard to read the scale.

In 1657 the Academy tested mercury as the liquid, comparing it with an identical bulb and tube containing spirit, and were disappointed to find that it did not expand as much as the spirit, although they did note that it responded more quickly to changes of temperature. Consequently they abandoned mercury, which, as Middleton says, 'was a pity'.

Boulliau in France and Boyle in England

The Florence thermometer spread rapidly, being used in Paris by the astronomer Ismaël Boulliau (also known as Bullialdus) to make systematic observations from May 1658 to August 1660, while another was brought to England in 1661 by Robert Southwell (later President of the Royal Society) who acquired it during his 'Grand Tour' of Europe. He showed it to Robert Boyle, who then had several made 'which are more conveniently shaped and more exact than the pattern I caused the first to be made' – although we do not know what these improvements were.

So by 1660 the spirit-in-glass thermometer had evolved to a reasonable state of development, although any degree of intercomparability was due to the skills of the workmen who made them rather than being due to reproducible calibration methods. For the next century, therefore, efforts were directed at improving accuracy and working towards standardisation.

Fixed points and scales

Christiaan Huygens

The freezing point of water was often used as one point when marking a scale on a thermometer tube, although whether this was a constant or not was argued about for decades in the seventeenth century. Sometimes just one point was used, often ice, with the scale being a simple fraction of the volume from that fixed point, it being tacitly assumed that the change of volume with temperature was linear (which for water it is not). In 1663 the Dutch physicist Christiaan Huygens proposed that the reference point could be *either* the freezing point *or* the boiling point of water, although he was not suggesting using the two points together. The Paris Observatory, built in the 1660s, had deep cellars, and for decades the nearly constant temperature in the cellars was used as a reference point in preference to ice. But in the last three decades of the seventeenth century there were a number of attempts to identify more reliable fixed, stable points that had a good physical foundation.

Bartolo, Eschinardi and Renaldini in Italy

The first person to suggest the use of snow and boiling water as two stable reference points was Sebastiano Bartolo in 1672, who said that these were 'everywhere immutable, for they depend on fixed causes'. He also said that ordinary temperatures, such as body heat and heat in cellars, were subject to slight changes. Bartolo divided the distance between the fixed points into 18 degrees. His contribution, however, seems to have gone largely unrecognised. Then, in 1680, another Italian, Francesco Eschinardi, published an account of how he had graduated a thermometer using the ice and boiling points of water. Most textbooks attribute this idea, however, to Carlo Renaldini and date it at 1694, Renaldini being one of the original Florentine academicians and by now an old man. He describes an experiment to calibrate a thermometer from the ice to the boiling point and Middleton (1966) believes that he used them in a better way than Eschinardi did.

Dalencé in Holland and de la Hire in France

Other fixed points were explored; in 1688 Joachim Dalencé, for example, used the temperature at which water freezes and the melting point of butter. Philippe de la Hire at the Paris Observatory, at the end of the seventeenth century, suggested ice and the temperature of the observatory's cellar, although the two were rather close together. In 1711, de la Hire's son, Gabriel, also did some experiments, placing one thermometer in the air and another in water out of doors in freezing weather.

During the five hours it took the water to freeze, the reading of the thermometer in the water/ice mixture was absolutely steady. When all the water was frozen, the reading started to drop further, but instead of recognising that this meant that a mixture of ice and water had a stable temperature, he wrongly concluded the opposite; probably his view was coloured because he wanted to refute the idea for personal reasons.

Newton, Patrick and Hauksbee in England

In England around 1700, Isaac Newton used linseed oil as the liquid and the temperature when water begins to freeze as the lower reference. He wrote that this temperature can be found by putting the thermometer in compressed snow as it is melting, seemingly recognising what Gabriel had missed about the stability and repeatability of the temperature of melting ice. Newton's upper temperature was blood heat. However, the paper in which he reported this was not known to be by him for some time, and no serious meteorological use was made of his thermometer. By the time that it was known who had written the paper, better ideas were already in circulation. In London, and closely associated with the newly formed Royal Society, Francis Hauksbee (the Younger) made thermometers and barometers commercially in his shop near the Society's offices, but his thermometers appear not to have been very precise. Their interest lies in the scale, which was marked as 'extreme hot' at 0°, 'temperate air' at 45° and 'extreme cold' at 90° with various steps between, such as 'frost' at 65°. However, this is identical to a scale developed by the instrument maker John Patrick, and it is likely that Hauksbee simply copied it. This is also sometimes known as the Royal Society scale, although it probably had nothing to do with them in reality – just a loose association because of close physical proximity.

By the start of the eighteenth century, it had been realised that it would be useful if meteorological observations made in one place could be compared with those made elsewhere. So the history of the thermometer throughout the century was mostly about competing scales, and we now look at the people concerned. By the end of the century, only three serious contenders remained – Réaumur, Celsius and Fahrenheit; but we will take a brief tour of all the contributors.

Rømer in Denmark

Meaningful intercomparisons of temperature recorded at different places could be achieved either by using thermometers made by the same maker and adjusted in the same way, or by agreeing on fixed points that could be reproduced anywhere using simple techniques. By the middle of the eighteenth century there were 18 different scales. I will not name them all, but instead will look at just the main players.

Ole Christensen Rømer, the famous Danish astronomer, who discovered that light had a finite speed, kept a notebook titled *Adversaria* (Latin for 'notebook') covering many topics and experiments. In the printed version of 1910, pages 202–13 deal with thermometers, in which he describes how he tested, using a drop of mercury, whether the bore of the tube was regular along its length before the bulb was blown. When the thermometer had been made, filled (with spirit-of-wine) and sealed, the positions of the column when placed in crushed ice and then in boiling water were marked as 7.5 and 60, respectively. It is estimated that he made these thermometers in 1702 and so was the first to make reproducible instruments using these two fixed points. Rømer probably chose 60 for the upper point because it was a number familiar to astronomers, and 7.5 as the lower (rather than zero) so that negative numbers did not often occur in meteorological use in Europe (7.5 is one-eighth of 60). Apart from being the first thermometers calibrated using two fixed reference points, they are of particular interest because Rømer was visited in 1708 by a young instrument maker, Daniel Gabriel Fahrenheit.

Fahrenheit in Poland and Holland

There is much, if rather inconclusive, literature concerning Fahrenheit's scale and to what extent his work was inspired by or was based on Rømer's. It is not necessary to go into all the details, which are not fully agreed anyway, but the outcome, however arrived at, is as follows: the spirit thermometers that Fahrenheit made around 1714 were calibrated on Rømer's scale (which had changed to the extent that freezing point was now 8° rather than 7.5°). But in 1713 Fahrenheit started making experimental thermometers using mercury as the liquid and began selling them in 1717. His scale was based on three fixed points. The lowest point has several explanations, the most probable being that it was the coldest temperature he could generate – using a mixture of ice, salt and water (used to make ice cream in those days). The upper reference was body temperature, but why it was 96 degrees (now it is 98.6°) and why freezing was 32 degrees is not clear, although there are about six different suggestions. Middleton (1966) does not speculate, and neither will I, about any of them. Fahrenheit did not use the boiling point of water as a reference, but on his scale it is 212°.

When Fahrenheit died in 1736, his work was continued by Hendrik Prins, who soon changed the upper reference point to the boiling point of water in place of body temperature. For meteorological use, the top end of such a scale is wasted, however, and most thermometers cover only up to about 150°. The scale quickly became adopted in the Low Countries (Belgium, the Netherlands and Luxembourg) and in the English-speaking nations, but was never adopted in France. Many people in the UK and America still use the Fahrenheit scale.

Réaumur in France

René Antoine Ferchault de Réaumur worked in many fields of the newly emerging sciences, including metallurgy and botany, but his work on thermometry is not his greatest achievement and he did not improve greatly on existing methods. He wrote extensively and left descriptions of how he calibrated his thermometers, the details of which can be found in Middleton (1966). Middleton is clearly irritated by Réaumur's verbosity and lengthy documents, in comparison with Fahrenheit's brevity (which left us, however, without an understanding of the derivation of his scale). It is quite difficult to summarise exactly what Réaumur did, but the following is as close as I can get.

Réaumur's scale is referenced to one fundamental temperature, that of melting ice, although it was still not certain whether this (or the boiling point of water) were really constant. The scale was based on equal volumes of expansion or contraction from the reference point. Réaumur used spirit-of-wine as the liquid and describes in considerable detail how he calibrated his thermometers (which we will not go into). It was not pure spirit-of-wine that he used, however, but spirit diluted with water. He recognised that different proportions of spirit-to-water gave different degrees of expansion as the temperature changed and so performed experiments using different dilutions, from which he determined the liquid's degree of expansion from the ice point to the boiling point *of the spirit-and-water mix* (not to that of boiling water). He finally chose a dilution for which the expansion was 80 parts in 1,000 (80 was used simply because it was a convenient number to divide into parts). But the public got the impression (understandably from his rambling and ambiguous notes) that 80 referred to the boiling point of water while it was really the boiling point of the spirit/water mix. Middleton felt that Réaumur was not too clear in his own mind about what he was doing. There was also uncertainty about what the effect might be of any air remaining dissolved in the liquid, or remaining in the tube above it. In France there was no immediate competition to the Réaumur scale and he had great authority there, preventing any criticism, but the scale gradually changed such that by 1770 what was still called Réaumur's scale was in fact somewhat different. Even as early as 1736 Réaumur had recognised that his methods were difficult to carry out by artisans and he was also talking about mercury in place of spirit. Such mercury thermometers were being made using Réaumur's scale and one was taken on an expedition to Lapland in 1736, the leader of the expedition later reporting that in January the mercury thermometer went down to 37° below zero, while the spirit-of-wine thermometer froze. Soon people wanted mercury thermometers corresponding to the Réaumur scale, one being produced by François Boissier de Sauvages, who gave instructions on how to calibrate a mercury thermometer in boiling water and then in snow or crushed ice, dividing the interval into 87 degrees ('so that it will agree with Réaumur's *spirit* thermometer'). It soon became common practice to define the melting point of ice as zero and the boiling point of water as 80° and to graduate the scale uniformly between them.

Delisle

In 1724, Joseph-Nicolas Delisle, a French astronomer, made some spirit thermometers with 0° indicating boiling point and 100° showing the temperature of the observatory cellars. Subsequently, while working in St Petersburg for Peter the Great, he made more but, finding it difficult to adjust spirit thermometers in boiling water, he adopted mercury as the fluid. Also lacking such deep cellars in St Petersburg, he used the one fixed temperature, that of boiling water, and scaled the thermometer in steps of 1/100,000 of the volume at boiling point. Later, after Josias Weitbrecht, following Delisle's suggestion, had measured the change in temperature of the water under the ice of the Neva River all winter and finding it did not change, Delisle adopted the two fixed points. Then, finding that the volume of mercury changed by about 150/10,000 between these two fixed points, he marked boiling point as 0° and freezing point as 150°. Weitbrecht appreciated that the expansion of the mercury depended also on the glass, since the mercury is contained in an instrument of variable capacity that changes with temperature. The Delisle scale made in this way was used in Russia for 50 years, but it represented little of value and did not persist.

'Centigrade' scales

Having ten fingers, we count in tens. There is nothing intrinsically special about a decimal base; it is not a basic law of physics. Counting in twelves has some advantages, because it has more divisors, and so we have used it for time measurement – 60 minutes, 12 and 24 hours – and for circular measurement we also use 30 and 60 as fundamental divisions. But since we do use ten as a base for everyday counting, and since 100 is consequently a 'round number', it was inevitable that sooner or later someone would devise a thermometer scale with 100 degrees between the fixed reference points. There is nothing particularly clever about the idea; it was simply an idea waiting to be expressed. But who devised the centigrade scale that we recognise today, with 0° representing the melting point of ice and 100° the boiling point of water? While Anders Celsius is now associated most widely with the scale, there were others, and we need to look at them too.

Delisle, Micheli, Horrebow and Réaumur

Delisle was probably the first to use such a scale (see above) when he made (spirit) thermometers with 0° indicating boiling point and 100° the temperature of the observatory cellars; this scale was never adopted.

The next known use of a scale of 100 derives from the work of Jacques Barthélemi Micheli in Geneva, who published a pamphlet in 1741 describing a spirit thermometer with zero representing the temperature of a cellar and 100 as the boiling point of water. This gave a temperature of –10.4° for melting ice.

Exchanges of letters between Peter Horrebow (Rømer's successor) and Réaumur show evidence that one of them suggested to the other a centigrade scale (100 as the ice point), but neither appears to have taken any practical action over making a thermometer with such a scale and it is still uncertain whose idea it was.

Celsius, Ekström, Strömer, Linnaeus and Christin

Professor of Astronomy at Uppsala University from 1730 until his early death in 1744, Anders Celsius appears to have become interested in thermometry around 1734, demonstrated by correspondence with Giovanni Bianchi in Rimini. In 1731 Celsius was operating a Hauksbee thermometer (and barometer) for official observations. He noted that he was also using a thermometer made by Abbé Nollet to Réaumur's design, along with one made by Delisle. One of the latter survives, kept at the Meteorological Institute of Uppsala University. Delisle's scale marked on the wooden base runs from 0° at the top, representing boiling point, with 150° for melting ice, the bottom of the scale being 205°. On this scale Celsius has marked his own, with 100 being opposite Delisle's 150 (actually opposite 151.8, probably due to a recalibration by Celsius). Celsius marks boiling point as 0°, opposite Delisle's 0°, noting that this was done at a barometric pressure of 25¼ Swedish inches. This marking was done on or before 25 December 1741 and regular observations were made using it from 2 June 1743, along with readings from thermometers by Hauksbee, Delisle and two by Hendrik Prins (Fahrenheit's successor). Shortly after Celsius's death, records from the Delisle and Prins instruments were discontinued, but a new one, labelled Ekström, was added.

Daniel Ekström was a leading instrument maker in Stockholm who had just returned from England, where he had been learning his craft. His scale had freezing point as 0° and boiling point as 100°. In September 1747, a thermometer bearing the name of Celsius's successor, Märten Strömer, was added, Hauksbee's having been removed earlier. Strömer's instrument had freezing point as 0° and boiling point as 100°. In the official record, Celsius's scale was replaced by Strömer's in April 1750, creating the impression that it was he who 'inverted' Celsius's scale.

There is some evidence that it might have been the botanist Carl von Linné (known as Linnaeus) who suggested the scale we now call the Celsius scale, but letters exchanged by me with Olof Beckman and Anders Moberg at Stockholm University, published in the Royal Meteorological Society's magazine *Weather* (Strangeways 2000), suggest otherwise. According to Beckman and Moberg (2000):

> It appears that the reversal of the original Celsius scale was more of an agreement among scientists in Uppsala and Stockholm at the time, rather than an initiative by any individual. From the scientific point of view the most important contribution to the modern temperature scale is due to Celsius because of his careful experiments on the fixed points.

At this time, Jean-Pierre Christin was working on thermometers in Lyon, and was secretary of the Académie des Sciences, Belles-Lettres et Arts de Lyon. According to Academy manuscripts, Christin's involvement probably started when he received a Réaumur thermometer calibrated at the Royal Observatory in Paris in 1736. He seems to have been completely wedded to the Réaumur scale, whether the liquid was spirit or mercury – the latter being his preference. In 1741 there is still no mention of a centigrade thermometer, but in 1745 Christin writes about an experiment in which he finds that 66 units of volume of mercury expanded to 67 units when the temperature was raised from that of ice to boiling point. From this he goes on to say, therefore, that 6,600 units would expand to 6,700 units. The difference, 100 units, is the number of degrees that he gave to his scale. Christin went on to comment that this has the advantage that just one of the two fixed points can be used to calibrate a thermometer because of the known change of volume with temperature. This is so much a Réaumurian way of thinking that it looks as if Christin wanted to use the centigrade scale, but wanted to remain faithful to Réaumurian logic in justifying it. This *thermomètre de Lyon* was used widely in Lyon and parts of the south of France, but Réaumur's reputation was so powerful that his thermometer became ubiquitous in France, ousting Christin's. Celsius was virtually unknown in France.

In conclusion, there seems little doubt that Celsius produced the first centigrade scale in 1741 with melting ice and boiling water as reference points (although others had used 100 steps but with different reference points) and that someone in Uppsala or Stockholm inverted it no later than 1745, to give today's Celsius scale.

The absolute, or thermodynamic, temperature scale

Guillaume Amontons

Robert Boyle and Guillaume Amontons had done experiments which demonstrated the relationship between temperature, pressure (or 'spring' or 'elasticity') and the volume of a gas. All the gases tested were found to have more or less exactly the same simple relationship in which the pressure increases linearly with temperature if the volume is kept constant (or the volume increases linearly if the pressure is kept constant). Even in the eighteenth century, referring temperature to that of melting ice or boiling water (or anything similar) seemed rather arbitrary and Amontons felt that there should be some form of *natural* temperature scale. He had done work in the 1680s on a complex three-liquid thermometer, derived from a three-liquid barometer developed by Robert Hooke, and argued that if you measure by how much the pressure of a gas falls as you cool it from boiling point to ice point (assuming the change is linear all the way down), it is simple to calculate what the temperature will be when the pressure is zero. His measurements showed that the pressure fell from 1 to 0.71 units when thus cooled, so that if cooling was continued down at this rate the temperature would be –248°C when the pressure reached zero. (Being French, he was of course using Réaumur's temperature scale, not that of Celsius). Lambert, Gay-Lussac, Flaugergues, Rudberg and others all did their own estimates of the change of pressure, each arriving at a slightly different value (probably because they did not dry the air completely). In 1847, based on the mean of four estimates of the change, Regnault arrived at a value of –272.75°C. In 1853 W.J.M. Rankine estimated zero to be –274.6°C. Increasingly precise experiments have set the accepted value today at –273.15 ± 0.02°C, this value being known as *absolute zero*. The early work was carried out before it was understood that a gas is composed of atoms and molecules in constant motion (Brownian movement), the motion creating the pressure, later explored and explained in 1827 by Robert Brown.

William Thompson

In 1848 the Scottish physicist William Thompson, later Baron Kelvin of Largs, proposed the *absolute* or *thermodynamic* scale of temperature, in which absolute zero corresponds to the temperature at which all molecular motion stops. As a tribute to him, the scale is known as the Kelvin scale, one degree Kelvin, known as a kelvin (with a small k), symbol K (a capital K, without the ° sign) is equal to one degree Celsius and absolute zero (0 K) corresponds to –273.15 on the Celsius scale. The melting point of ice is thus +273.15 K and the boiling point of water is +373.15 K. However, this scale is used mainly for scientific purposes.

The International Practical Temperature Scale

The scale of temperature used today is the *International Practical Temperature Scale* (IPTS) agreed in 1968. It is based on a number of equilibrium states, including the triple point of water and the boiling point of water (see discussion of calibration below) and also the boiling point of oxygen (−182.962°C or 90.188 K) at standard atmospheric pressure. There are additional secondary reference points, one of which is the familiar ice point of water (0°C). Although called the IPTS it is, in practice, really just the familiar Celsius scale.

Modern thermometers

Liquid-in-glass

'Present temperature'

The most commonly used method of measuring temperature today is still the thermal expansion of a liquid in a glass tube as pioneered by Fahrenheit and Celsius; little has changed in their design.

Until the 1970s, in the UK, temperatures were measured to the nearest 1°F. Since then a change has been made to estimating the reading to the nearest 0.1°C from thermometers graduated in 0.5°C steps (Figure 2.3).

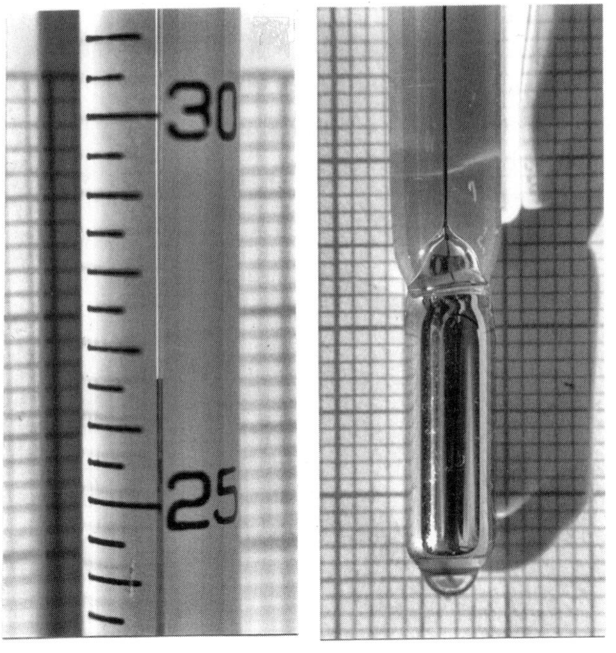

Figure 2.3 A modern mercury-in-glass thermometer, showing the bulb and the division of the scale into half-degree steps, estimated by eye to tenths of a degree, in this case 26.7°C. The glass bulb contracts slowly with time, increasing the readings slightly, hence the importance of regular calibration for climate applications.

Meteorological thermometers are manufactured to within certain tolerances. For mercury, from −40 to 0°C the maximum error is typically between −0.3 and +0.2°C, while from 0°C upwards it is between −0.2 and +0.05°C. For spirit thermometers, below −40°C the maximum error is ±0.06°C, from −40 to 0°C and from +25°C upwards it is ±0.25°C, while from 0 to +25°C it is ±0.1°C.

Readings may be in error anywhere within these limits. The figures are typical UK values and will probably vary from country to country and from manufacturer to manufacturer. Some thermometers are supplied with calibration certificates, but if they are not, it is necessary to check the zero before use for climate work (see below under the heading *Zero drift*). It needs to be remembered that these thermometers were made for early and mid twentieth century basic meteorological use, not for following small climatic changes. This is an important point to remember.

The second type of error that the Met Office describes arises if the liquid column is not read by the observer at eye level, a parallax error being introduced which can be up to ±0.2°C. These sorts of errors, however, will probably not have a systematic bias in one direction at all stations and at all times, and so are less of a problem than errors that are always biased one way.

A mercury-in-glass thermometer has a *time constant* of about one minute (the time to reach 63% of the final value after a step change in the temperature) for a bulb of 10 mm diameter in an airstream of 5 m s^{-1}. In practice, temperatures tend to be cyclic and uneven and it is difficult to predict theoretically how a thermometer will behave under such conditions, but any temperature fluctuations faster than the time constant will not be registered. When taking a reading of the 'present temperature' it is desirable that the thermometer does not respond too quickly, as it would then be unduly affected by short-term variations. The average value over a period of about 5–10 minutes is of the right order, and for this a time constant of not less than about one minute is appropriate.

Maximum and minimum thermometers

The idea of designing thermometers that record the maximum and minimum temperatures was first suggested by Jean Bernoulli in 1698, but his ideas were based on air thermometers and were not very practicable. There followed dozens of designs using liquid-in-glass, again most of them impracticable. Then in 1790 Daniel Rutherford described to the Royal Society of Edinburgh a minimum thermometer, invented by his namesake John Rutherford. This is, in principle, exactly the same as those used by most National Meteorological Services (NMSs) today. It uses spirit as the liquid, in which a slide-indicator or index is free to move along the bore but which stops when it gets to the liquid's meniscus (Figure 2.4). The thermometer is set by inverting it so that the indicator falls and rests on the meniscus; it is then laid on its side. As the temperature rises, the slide stays where it is, with

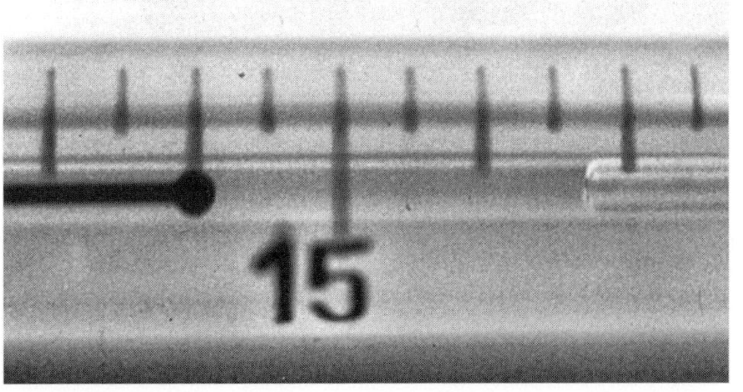

Figure 2.4 As the temperature falls, the meniscus of the spirit pulls the movable index down with it. If the temperature later rises, the index stays put, with the liquid going round it, recording the minimum temperature reached. It is reset by turning the tube upside down so that the index falls back to the meniscus. Both the maximum and minimum thermometers are operated horizontally to remove any influence of gravity on the index or mercury column.

the liquid going round it. When it falls, the meniscus pulls the slide back towards the bulb, the slide being unable to break through the meniscus owing to surface tension. The minimum temperature is where the top of the slider rests. The spirit, which is an organic compound, changes under exposure to light, when polymerisation occurs, causing the volume of the liquid to reduce, thus introducing an error that is difficult to quantify.

Initially, attempts were made to use the same principle (of a movable index), in this case above the mercury column, to record maximum temperatures, but they failed because the mercury could get round the indexes. Despite many attempts to solve the problem, none were successful. While several alternatives were tried, it was the Negretti and Zambra design, patented in 1852, that finally proved the best and is the method still in use today. These thermometers are identical to ordinary mercury-in-glass thermometers, except that there is a constriction in the mercury column between the bulb and the lowest graduation on the stem (Figure 2.5). As the temperature falls, the mercury column breaks at the constriction and is not drawn back into the bulb, thereby holding its position. If the temperature rises again and to a higher value, the column is reformed and the reading rises to the new high. It is then held at this new point when the temperature again falls. To prevent gravity from moving the detached mercury column, the thermometer is operated on its side. It is reset by a sharp shake downwards, as with an old-fashioned clinical thermometer. Tolerances of manufacture, errors and resolution are the same as for the ordinary type of thermometer.

In 1782, before either of the above methods was developed, James Six, of Canterbury, designed a combination instrument for maximum and minimum readings. However, it used both mercury and spirit, and it was found that the spirit

Figure 2.5 As the temperature increases, the mercury is forced through the narrow constriction. If it later cools, the mercury column breaks at the constriction, leaving the top of the column at its highest point, recording the maximum temperature. It is reset by shaking, like an old-style clinical thermometer.

could get round the side of the mercury. It also had the problem that the scale was on the backing-board onto which the thermometer was fixed, and this prevented accurate readings being made and also risked movement of the thermometer on the plate. It is no longer used for meteorological work, but many greenhouses, both professional and amateur, have one.

Maximum and minimum readings together are used widely to derive an approximate daily mean temperature by taking 1/2(max + min) and this is discussed further in Chapter 8. They also give an indication of the daily range of temperature – the *diurnal temperature range* (DTR). Both are considered further in later chapters.

Zero drift

It was assumed that once a liquid had been sealed into the thermometer, stability would result. There were still a few remaining doubts about the permanence of spirit-of-wine, with some question as to whether air that was dissolved in the spirit might collect as bubbles and increase the volume. But when mercury became the favoured liquid even that uncertainty was removed. However, doubts returned in 1789 when Jacques-Dominique Cassini de Thury, following tests in the cellars of the Paris Observatory, noticed that there seemed to be signs of a very slow rise combined with the very small annual variations in temperature. He suspected 'some peculiar cause independent of the state of the external atmosphere', but he followed it up no further and it remained a vague unanswered question.

The history of how this matter was resolved is complex and we will not go into all the details (see Middleton 1966 for the finer points). What evolved was the recognition that there was a slow, small rise of the zero (and so of all the readings) over time, going on for many years. In 1873 James Prescott Joule reported observing the gradual rise of the ice point of two thermometers that he had checked over 29 years, the rate of the rise slowly decreasing over time.

By the late 1880s the cause had been narrowed down to the contraction of the glass bulbs of the thermometers, and this was finally identified as being due to the slow rearrangement of the molecules of the glass. The composition of the glass was also thought to be important, with 'hard glass' (containing less lead oxide) being preferred. The Met Office handbook today gives a figure of 0.1°C per decade initially, 'even with the best glass'. Thus thermometers used for several decades, if not recalibrated regularly (see later), will give the impression of a small gradual temperature increase.

Since the potential rise of temperature due to the contraction of the bulb is of the same order of magnitude as the observed climatic temperature rise, it very well illustrates the difficulties of using simple instruments, intended for less demanding purposes, to detect very small climatic changes. There will inevitably be this sort of error in the datasets and, being gradual, the changes will be indistinguishable from natural changes and so not easily detected. It seems surprising that a matter as simple as contracting glass could conceivably explain some of the apparent observed rise in temperature. This characteristic is, nevertheless, an established fact and one of the two main drawbacks of liquid-in-glass thermometers.

Bimetallic thermographs

These instruments are mentioned here for completeness, but they are of relatively low accuracy and thus of no use in climatology. The sensor is a bimetallic strip coiled into a helix. One end of the helix is fixed to the instrument's case, in such a way that the zero point can be adjusted by rotating the helix. The temperature range is adjustable by altering the length of the helix. The other end of the helix is free to rotate, moving an arm holding a pen across a paper chart. (A bimetallic strip is a composite of two metals with different expansion rates welded together. As the temperature changes, the strip bends.) A spring- or battery-operated drum rotates the chart. A thermograph can, at best, have an accuracy of ±0.5°C but, to achieve this, care is necessary in making the initial settings and any necessary periodic adjustments. In practice it is possible that the readings may be out by considerably more than 0.5°C.

Electrical thermometers

Platinum resistance thermometers

Even though Sir Humphrey Davy announced his discovery that the electrical resistance of metals is dependent on temperature in 1821, his proposal to use this characteristic for thermometry was not taken up for another 50 years, when Carl Wilhelm Siemens discussed its use for observing temperatures in inaccessible locations (Siemens 1871). Although intended for industrial applications, Siemens suggested that it could also be used in meteorology, but he took the matter no further and it fell to Hugh Longbourne Callendar to investigate further. He carried out elaborate experiments relating the readings of an air thermometer to the resistance of a platinum resistance thermometer (PRT). The first meteorological use of PRTs was made at Kew Observatory in 1896.

Various metals and alloys can be used, the choice depending on such factors as the required stability over a long period, the uniformity of the product, the degree of change in resistance with temperature, and cost. All the desirable features do not occur together in one metal, but the most commonly used is platinum because of its stability and relative linearity; nickel is often a second choice when cheapness is important. Although all metals have a nearly straight-line temperature response, they are not completely linear and for precise work a correction may be necessary.

Sensitivity to temperature change is defined by the *temperature coefficient* of the material. For platinum this is $3.9 \times 10^{-3}\,K^{-1}$ while for nickel it is $6.8 \times 10^{-3}\,K^{-1}$. Taking the example of a *platinum resistance thermometer* (PRT) (Figure 2.6) with a resistance of 100 ohms at 0°C (the usual standard today and the most-used metal), the increase in resistance for a 10-degree rise is $100 \times 3.9 \times 10^{-3} \times 10$ (that is, 3.9 ohms), giving a total resistance at +10°C of 103.9 ohms, and so on, proportionally, for increasing and decreasing temperatures. 'Grade A' PRTs have a manufacturing tolerance of ±0.15°C from −50 to 0°C, increasing to ±0.25°C at 100°C.

To measure the change in resistance, the thermometer can be made one arm of a Wheatstone bridge, or a constant current can be passed through the sensor, the voltage across it being measured. In both methods, current flows during the measurement, thereby heating the sensor. Provided that the current is kept low, that the readings are made quickly and that there is an adequate flow of air past the sensor, the heating effect can be ignored. When a data logger is used to record the resistance, readings are taken in milliseconds and so heating is negligible.

Although resistance thermometers are now mostly used in automatic weather stations, they have also been used at manual stations, with a Wheatstone bridge being balanced by the operator. This takes time and so self-heating is more of a concern, but the advantage is that the temperature can be read remotely without having to go to the screen and open it (see next chapter).

Figure 2.6 Three thermistors and a platinum resistance thermometer alongside a pin to indicate scale.

When the sensor is situated some distance from the measuring equipment, be it manual or automatic, it is necessary to compensate for the resistance of the wires between the sensor and the bridge, which would otherwise unbalance the bridge as well as introduce unwanted additional resistance changes due to temperature. Compensation can be achieved by including extra leads at the opposite side of the bridge to balance out those of the sensor or, alternatively, the current can be fed along one pair of wires while the signal voltage across the sensor is picked off directly by another pair of wires that carry no current. A further complication of the low resistance of PRTs is that contact resistance within plugs and sockets can become a source of error.

It is possible to use software in a data logger to detect and store maximum and minimum temperatures as well as average or instantaneous values, the one PRT supplying all the readings.

Thermistors

Thermistors are semiconductor heat-sensitive resistors that have a temperature coefficient an order of magnitude greater than those of pure metals. Sensors with both positive and negative slopes can be made, and all types have a much higher value of resistance than PRTs. In addition they can be made very small (Figure 2.6), which is useful in certain applications requiring rapid response times or minimal radiative heating. The higher resistance of the sensors compared with a PRT (typically ranging from 3,000 to 100,000 ohms at 20°C) eases the problem of lead resistance and of plug and socket contact resistance.

The non-linear response of thermistors is a slight disadvantage, but this can be corrected by the use of two- or three-element composite thermistors within the probe. Alternatively, corrections can be made in the logger or when processing the readings. Depending on how the linearisation is done and how the corrections are made, the accuracy of a thermistor varies from about ±0.05 to ±0.5°C. An aging process occurs and, to reduce this, thermistors are temperature-cycled by the manufacturer, keeping the long-term drift typically within the range of ±0.015°C per year. This, however, is still of the same magnitude as caused by the shrinkage of the glass bulbs of mercury-in-glass thermometers and so recalibration is necessary periodically; platinum resistance thermometers are much more stable.

Cheaper thermistors may vary in resistance between one and another, and so are not directly interchangeable without recalibration. (Curve-matched sensors are, however, available at little extra cost, and these can be directly interchanged. But they still need periodic recalibration.) The same problem of non-interchangeability occurs with the cheaper PRTs.

Thermocouples

In 1822, T.J. Seebeck showed that when two different metals touch, a small voltage is generated. If two such junctions are formed and one is kept at a temperature different from the other, the consequent difference in potential between the junctions is proportional to their difference in temperature. This is known as a *thermocouple* and can be used to measure temperature *difference*. If a number of such junctions are connected in series, the voltages are added together and the combination is known as a *thermopile*.

Because thermocouples measure differences in temperature rather than absolute temperature, they are well suited to applications such as radiation sensors (see Chapter 1). To measure absolute air temperature with a thermocouple, however, it is necessary to hold one set of junctions at a known temperature (for example 0°C) and this is not very practical in the field. Thermocouples also produce rather low output voltages and so require higher amplification than the resistance types of temperature sensor.

Thermal capacitors

Intended for use in radiosondes, these very small sensors have a short time constant. They are discussed further in Chapter 10.

Satellite measurements of surface temperature

Measurements of many meteorological surface variables, including temperature, have begun to be made from satellites over the last 30 years. Whereas this will be useful in the future, such measurements account for very little of the past or present data (but see Chapter 10 with regard to satellite measurements of temperature profiles through the depth of the atmosphere using sounders). The techniques for measuring variables such as surface air temperature or sea surface temperature from space still have many limitations and are in no way a substitute for ground-based observations. However, for completeness, something needs to be said about remote sensing, and Appendix D outlines the basic principles.

References

Beckman, O. and Moberg, A. (2000) The Celsius thermometer. *Weather* **55**, 392–4.

Fludd, R. (1626) *Meteorologica Cosmica* (*Philosophia sacra et vere Christiana seu Meteorologia cosmica*). J. T. de Bry: Frankfurt.

Middleton, W. E. K. (1964) *The History of the Barometer*. The Johns Hopkins Press: Baltimore, MD.

Middleton, W. E. K. (1966) *A History of the Thermometer and its Uses in Meteorology*. The Johns Hopkins Press: Baltimore, MD.

Middleton, W. E. K. (1994) *The History of the Barometer* (reprint). Baros Books: Trowbridge, UK.

Siemens, C. W. (1871) On an electric resistance thermometer for observing temperatures at inaccessible locations. *Proceedings of the Royal Society* **19**, 443–5.

Strangeways, I. C. (2000) More on temperature scales. *Weather* **55**, 145–6.

Strangeways, I. C. (2003) *Measuring the Natural Environment* (2nd edn). Cambridge University Press.

3

Screens, stands and shelters

The need for thermometer protection

If a thermometer (of any type) is placed in a liquid, the exchange of heat between the two proceeds by conduction, convection (usually stirring) and radiation, the first two being the primary influences, radiation being almost negligible. In air, conduction plays only a very small part, convection and radiation bringing the thermometer into equilibrium. Unlike most liquids, however, air is very transparent to infrared (IR) radiation so that the radiative exchange is between the thermometer and any solid or liquid bodies that are in visual range of the thermometer, however far away, including the sky. To this long-wave radiation must be added the short-wave radiation directly from the Sun, reflected sunlight, including that from clouds, and diffuse scattered radiation from the sky. If the exchange was by radiation alone, the thermometer would read a mean of the temperature of every object around it, not that of the air. There are two ways in which the thermometer can be made to read just the temperature of the air, either by ensuring that every object 'seen' by the thermometer is at air temperature, or by increasing the exchange of heat by convection to the extent that radiation becomes negligible.

Only in the case of very small thermistors, platinum resistance thermometers (PRTs) made of fine wire, thermocouples or thermal capacitors will radiative heating be negligible. These are used in radiosondes and are considered in Chapter 10.

The evolution of screens

It was not at first realised that thermometers had to be protected from solar radiation and rain, and particularly from the less obvious sources of infrared radiation (see Chapter 1 concerning Count Rumford's discovery of IR radiation in 1804). As soon as this became clear, thermometers began to be installed on north-facing walls or even in unheated north-facing rooms. However, the Reverend Henry Miles of

Figure 3.1 Thatched tropical shelter at Hong Kong Royal Observatory (photograph kindly supplied by Joan Kenworthy).

Tooting wrote a paper (Miles 1747) describing how two identical thermometers, one exposed in an unheated room and the other outside the window only a few feet away, displayed very different temperatures. This ended the idea of indoor exposure.

Open shelters

Seventy years later, Fourier gave a detailed explanation of the effects of radiation on a thermometer, showing how a thermometer could be either warmed or cooled by radiation. Ideas were now developed about how to protect thermometers from radiation and this led to the design of many structures. The name *screen* was used in England and in its Commonwealth to refer to structures that protected the thermometers from radiation coming from *all* directions (except from the screen itself of course). Those that provided protection only from rain and direct solar radiation were called *shelters* or *stands*.

George Martin

One of the first stands was made by George Martin on the Isle of Wight around 1838, consisting of a thatched roof supported on wooden posts, with a small chimney to prevent warm air from accumulating under the roof. Larger versions became very common in the tropics (Figure 3.1). However, because the thermometers were fixed at the lowest part of the roof, they received radiation from the ground outside.

Figure 3.2 The Lawson stand (illustration kindly supplied by the Met Office).

James Glaisher

Glaisher's stand, attributed to James Glaisher but probably designed by the Astronomer Royal, Sir George Airy, was used at the Royal Observatory at Greenwich from 1841 and is more correctly called the *Greenwich stand*. But, as it had to be rotated during the day to keep the thermometers in the shade, it was rather impracticable, and so is not illustrated.

Henry Lawson and the Reverend Stow

Lawson's stand (Figure 3.2) was similar to the Greenwich design but did not have to be rotated. Also of fixed design was the stand designed by the Rev. F. W. Stow, but his field tests convinced him that louvred screens were better.

H. Marié-Davy and Charles Sainte-Claire Deville

The French initially took a different path, preferring large open stands. For example, while he was Director of the Observatoire de Paris, Hippolyte Marié-Davy started comparative observations in a large free-standing shelter on the lawn. It was 3 m high and 1 m square, with a sloping roof covered in green-painted cloth.

He was later Director at the new Montsouris Observatory, where he introduced this design. A subsequent Director of the Observatory, Charles Sainte-Claire Deville, designed a larger version of the Marié-Davy shelter, standing 7 m high, made of cast iron with 'wings' to the east and west to shade the thermometers in the morning and evening, and with evergreen shrubs planted to the east, south and west.

Fully enclosed screens

All of the early shelters had been open in some direction. While this allowed good ventilation, their openness also allowed radiation to enter. Screens were next designed that were closed all round, with ventilation being through louvres.

The Plymouth Dockyard screen

As usual, the start of an idea is hard to pin down but Middleton (1966) reports that the first documented account he was able to find concerns a screen at Plymouth Dockyard around 1835 which looks as if it was designed by a mariner who had travelled to the East. It is still open to the north but the surrounds are slatted, Venetian blind style.

The Kew stand

Again its origins are obscure, but the *Kew stand*, as it became known in the 1870s, apparently originated in Canada and was designed in 1845 by Sir John Henry Lefroy, Director of the Toronto Magnetic and Meteorological Observatory, based on a meat-safe design consisting of a louvred box with a second louvred box 18 inches (45 cm) outside the first. The Kew stand was much larger: indeed it was more like a ventilated shed.

The Stevenson screen

While Thomas Stevenson is often viewed as the originator of the louvred screen, he did not get involved until 1866, after the Kew development, although he was the first to use double louvres (Figure 3.3). At first the screen was small, but a larger model was developed (on the Royal Meteorological Society's recommendation) with a double roof and a floor made of overlapping boards above and below the sills. It was to be painted with two coats of white lead and one of zinc white enamel. Middleton (1966) comments, however, that while such double louvres made it impossible for radiation to enter (or leave) the screen without encountering solid matter, this advantage was at the expense of reducing air flow past the thermometers. He goes on to say that this is seen as an insuperable disadvantage, especially in hot countries.

Figure 3.3 Stevenson screen (at the Centre for Ecology and Hydrology, Wallingford, UK).

Standard Stevenson screens in the UK currently have internal dimensions of about 1 m width, 0.5 m height and 0.3 m depth, which is large enough to contain a variety of thermometers as well as a bimetallic thermograph and a hygrograph (Strangeways 2003). Figure 3.4 shows a typical standard thermometer set in a Stevenson screen. There are larger models of screen, and a half-width model is used to house just the liquid-in-glass thermometers. However, worldwide, there are a great many designs of louvred screens of the Stevenson type, some of them bigger than those described above (Figure 3.5). A few now have blackened interiors.

The cotton region shelter

Very similar to the Stevenson screen is the cotton region shelter (Figure 3.6). I have seen these in common use in Indonesia.

Screens on ships

How air temperature is measured over the sea from ships and buoys is discussed in Chapter 5; it is sufficient here to say that it is very difficult to measure air temperature on board ship for reasons that will become apparent later. The type of screen used is usually a miniature form of the Stevenson screen holding just two thermometers (dry and wet bulbs) and with single louvred walls (Figure 3.7).

Figure 3.4 The usual complement of liquid-in-glass thermometers in most screens comprises: a 'present temperature' mercury-in-glass thermometer (*left*); a similar mercury-in-glass thermometer with a wetted wick around its bulb for deriving relative humidity (*right*); a maximum thermometer (*top*) and a minimum thermometer (*bottom*).

Heinrich Wild

Heinrich Wild, as the young new Director of the Berne Observatory, devised a metal shield (Figure 3.8a) made up of two concentric cylinders with a 2-cm air gap between and a conical roof, this being hung on a north-facing wall 3 m above the ground. In 1868, Wild then moved to St Petersburg, where he became Director of the Central Physical Observatory and so was in charge of the large Russian Weather Service. In 1869 he designed a metal louvred screen to be hung on a north wall. But Wild then decided that a free-standing screen was better than one fixed to a wall, and he designed an extremely large construction 3 m high (Figure 3.8b). Inside this he placed his cylindrical metal screen. It was used at the more important Russian stations. We know nothing about the effects of snow on it (see later).

Signal Service thermometer shelter

Henry A. Hazen of the US Signal Service then wrote a paper (Hazen 1885) saying that the air in the large Wild screen was not at the same temperature as that outside, and a bitter argument started. Hazen's own experiments produced the Signal Service thermometer shelter and this was used for a long time by the US Weather Bureau. It was large with single louvres all round, the bottom being of solid wood; the roof was double with a 6-inch (15 cm) gap between the layers. Hazen believed (with good reason according to Middleton) that the Stevenson screen was much too small.

Figure 3.5 A typical larger screen of the Stevenson type (Tulkarm, West Bank).

Aspirated screens

The Stevenson screen was not immediately accepted in the rest of Europe and in France, in particular, the 'whirling' or 'sling' thermometer was considered better (Figure 3.9). The view was that it was more effective to increase convection rather than reduce radiation, which is true. However, the sling thermometer has not been widely adopted, except for the measurement of humidity – there being two thermometers on the sling, with one having a wetted wick around it (see below). The obvious reason for it not being taken up was that it needed an operator to turn the handle and this did not allow for continuous operation.

The Stevenson screen is ventilated naturally, relying on the wind to move air through it, but its double louvres hamper air flow and render it unsuitable for warm places and calm days. In Scotland, at about the same time as Hazen's work, John

Figure 3.6 A cotton region shelter (in Java, Indonesia).

Aitken (well known for his work on cloud droplet nuclei) had turned his attention to the exposure of thermometers. In his first paper on the topic (Aitken 1883), he pointed out the unsuitability of the Stevenson screen in warm climates. He designed several simple screens, these always giving considerably lower maximum temperatures than the Stevenson screen. He also did some significant experiments with an aspirated screen consisting of two coaxial tubes and concluded that ventilation by a suction fan was better than any other method and much better than a Stevenson screen in calm sunny weather (Aitken 1883). In 1887, Richard Assmann reported his first experiments with aspirated thermometers, which were very similar to Aitken's design, which resulted in the Assmann psychrometer, now used as a standard reference but not for routine measurements in the field: like the sling thermometer, it is not amenable to continuous operation. The psychrometer uses a clockwork fan which draws air at around $2\,m\,s^{-1}$ past two mercury-in-glass thermometers, one with a wetted wick around it, the instrument being used to measure relative humidity (RH).

Figure 3.7 A typical ship's small screen (aboard the British Antarctic Survey research ship *Bransfield* in the South Atlantic).

Miniature screens for automatic weather stations (AWSs)

In the early days of modern AWSs (the 1960s) electrical thermometers, usually PRTs, were sometimes exposed in wooden Stevenson screens. While this had the advantage of some continuity with manual systems, it was no longer necessary to use such large screens now that the sensors were smaller and did not need to be read manually, and so the first miniature temperature screens were developed. One of my first screen designs for AWSs in the mid-1960s (Figure 3.10) was very similar to John Aitken's (although I had not seen his designs at the time). A typical AWS screen of today is shown in Figure 3.11.

Height of screens

Because quite considerable temperature gradients can occur near to the ground, it is important that screens are mounted at the standard height of 1.25 m. Middleton (1966) remarks that the need to use a standard height does not seem to have been widely understood. Indeed the fact that air temperature varies considerably with height makes it very important that a standard height is used. The temperature varies with height because the atmosphere is heated mostly from the ground up, as explained in Chapter 1, the profile of temperature depending on the extent and

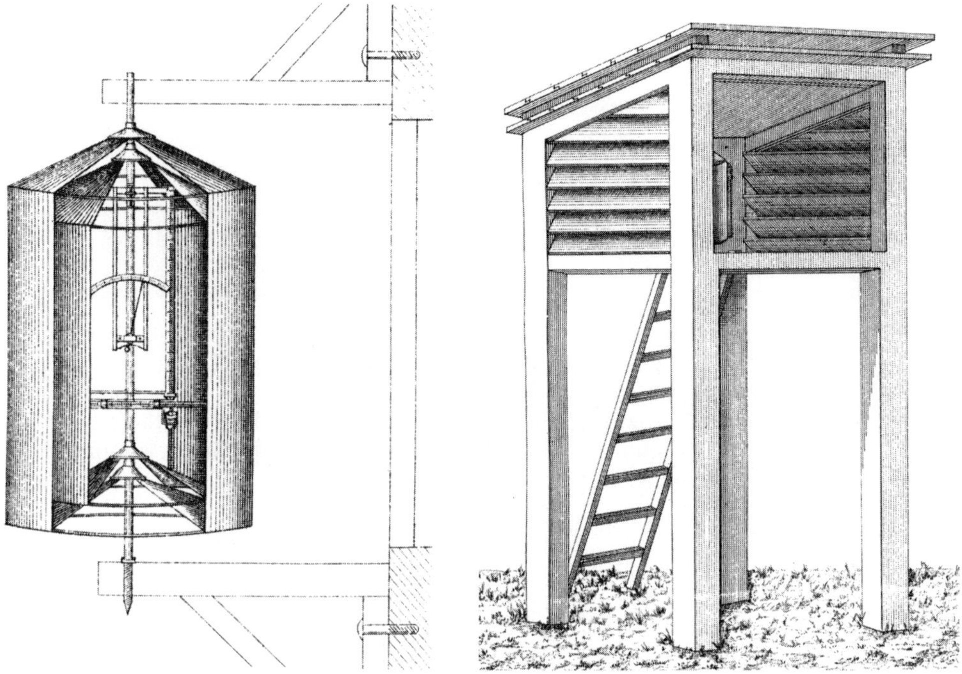

(a) (b)

Figure 3.8 Wild's cylindrical zinc shield, which was operated first on a north-facing wall (a) but later inside a very large screen (b) (drawings kindly supplied by the Met Office).

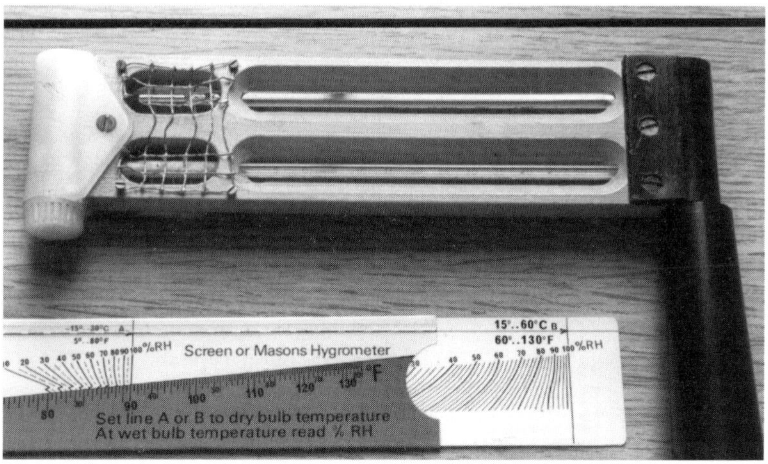

Figure 3.9 'Whirling', or 'sling', thermometer holding wet and dry bulb thermometers. The slide-rule allows the relative humidity to be calculated from the two readings.

Figure 3.10 An experimental thermometer screen which I developed for use on early AWSs while at the Centre for Ecology and Hydrology (CEH) (then the Institute of Hydrology; IH). The container beneath the sensors holds water to supply the wet-bulb thermometer to calculate humidity. At this time there were no reliable electronic humidity sensors (nowadays, capacitive sensors, originally developed for use in radiosondes, are often used).

Figure 3.11 Present-day AWS screens are mostly of the type illustrated here. They do not need water supplies as there are now reliable, accurate, small humidity sensors. Some now have blackened interiors.

nature of mixing; that is, whether mixing is brought about by turbulent air flow through wind action or by convection in the form of bubbles of rising warmed air. It could be argued that land air temperature is not clearly defined because of this variation with height, but provided that we recognise the importance of a standard height, a benchmark can be set. Where a standard height is not possible, the actual height must be known so that adjustments can be made. Looking at the great differences in height of Stevenson-type screens around the world, intercomparisons may not be very accurate. Over a snowpack that changes depth with season, the screen should be moved up and down to keep it the same distance above the snow.

However, it will be shown in later chapters that it is *changes* in temperature that climatologists are mostly interested in, not absolutes, so provided that the height of the screen is not changed during its lifetime, height is perhaps less critical from the climatological perspective.

Continuity of records

The problem of changing from one type of instrument to a new one is that it can introduce a step change in the observations, and this occurs if a change is made from Stevenson screens with liquid-in-glass thermometers to AWS screens with electrical resistance thermometers. The comparison of screens is thus important.

Comparisons of screen performance

A difficulty in comparing screens is deciding which, if any, gives the correct reading – that is, which measures most closely the 'true' temperature of the air outside. One way of making this decision is to assume that the screen giving the lowest maximum, especially on calm summer days, and the highest minimum, especially on calm clear nights, is nearest to the actual. It is also necessary, when comparing screens, to ensure that the heights of the screens above the ground are all the same, for the reasons explained above. It is also necessary to use identical electrical thermometers in all of the screens under test. These thermometers must also have been calibrated together and intercompared.

McTaggart-Cowan and McKay (1976) tested 19 screen designs, including aspirated models, and while they found that the aspirated screens all agreed to within ±0.2°C of each other, the non-aspirated screens were found to be less accurate, particularly with regard to maximum and minimum temperatures – each one being different.

Painter (1977) carried out intercomparisons in the UK and found that if the conditions were unfavourable (bright sunlight and low wind speed) the temperature measured in a Stevenson screen could be in error by +2.5°C, while on a cloudless calm night the within-screen temperature could be colder by up to −0.5°C.

It is important also to remember that if the screen is not kept clean it will be less effective, and that repainting should be done every two years. Keil (1996) showed that temperatures could be lowered by 1°C by repainting an old screen at a site in the UK. These figures are likely to be even higher in climates which have high solar radiation intensities.

Huband (1990) investigated the performance of three similar screens of the type used in modern AWSs (Figure 3.11), comparing them with each other and with a Stevenson screen. He found that measurements within the AWS screens agreed with those made inside the Stevenson screen to within about ±0.5°C, and for much of the time to within ±0.2°C, although the difference could at times be as much as ±1°C, the latter probably being due to the different time constants of the screens, since these larger differences occurred mostly at times of rapid temperature change. Maximum temperatures occurred within 10 minutes of each other, but there was more delay in the Stevenson's screen minimums, owing to the screen's greater thermal mass.

The UK Met Office did tests comparing several designs of Stevenson and AWS screens and found that none had as good a response as the aspirated screens that were used as the reference (Met Office 1999). Later tests (Met Office 2001) compared various wooden and plastic Stevenson screens, some of the latter having blackened interiors. All screens registered similar minimums, while the white plastic screen showed slightly higher maximums, the internally blackened screens, marginally lower (about 0.15°C) in high solar radiation. The efficacy of black interiors has yet to be established.

For precise climatological measurements, aspiration is essential. The importance of this will be returned to in Chapter 11.

References

Aitken, J. (1883) *Proceedings of the Royal Society, Edinburgh* **12**, 667.

Hazen, H. A. (1885) *Thermometer Exposure*. Professional Papers of the Signal Service, No 18. US Signal Service: Washington DC.

Huband, N. D. S. (1990) *Temperature and Humidity Measurements on Automatic Weather Stations: A Comparison of Radiation Shields*. Internal Report, Campbell Scientific Ltd.

Keil, M. (1996) *Temperature Measurement in a Stevenson Screen*. Department of Meteorology Report, University of Reading.

McTaggart-Cowan, J. D. and McKay, D. J. (1976) *Radiation Shields: An Intercomparison*. Unpublished report, Canadian Atmospheric Environment Service.

Met Office (1999) *Field Trial of Temperature Screens* (Bridgman, D. J., ed). Observations-(Logistics and Automation) Met O (OLA), Technical Report No 10 (Issue 2).

Met Office (2001) *Field Trial of Metspec Screens* (Sparks, W., ed). Technical Report: Document Identifier TR19, Issue 1.0 Original Issue August 2001.

Middleton, W. E. K. (1966) *A History of the Thermometer and its Uses in Meteorology*. The Johns Hopkins Press: Baltimore, MD.

Miles, H. (1747) Concerning the difference of the degrees of cold marked by a thermometer kept within doors, or without in the open air. *Philosophical Transactions* **48**, 613–16.

Painter, H. E. (1977) *An Analysis of the Difference between Dry-bulb Temperatures Obtained from an Aspirated Psychrometer and those from a Naturally Ventilated Large Thermometer Screen at Kew Observatory.* Met Office unpublished report. Available at the National Meteorological Library, Exeter, UK.

Strangeways, I. C. (2003) *Measuring the Natural Environment* (2nd edn). Cambridge University Press.

4

Measuring land surface air temperature

The origin of data

However mathematically cunning or appealing an analysis technique may be, any results or conclusions are dependent upon the quality of the basic data. In climatology, quality of the data is essential (Jones et al. 1999).

This should be kept in mind as we look at how temperature data have been collected from over the land and over the oceans.

If we wish to investigate how the global land surface air temperature has changed over the last 150 or so years, we have to use data collected by the old manual instruments described in the previous two chapters. Only with the arrival of automatic weather stations (AWSs) in the 1960s was there an alternative to these traditional instruments, but even today most data are still being collected by the age-old methods (AWSs are discussed in Chapter 11).

While innumerable temperature measurements have been made by countless organisations over the last 150 years, certain criteria must be met if the data are to be of any use for a precise study of climate change. One of the decisive factors is that there must be a sufficiently long run of uninterrupted data collection (preferably at least 30 years, although less is sometimes acceptable). Certain operational standards must also be met, on which more will be said later. It is mainly, but not exclusively, data collected by the world's National Meteorological Services (NMSs) that come anywhere near to (potentially) meeting these requirements, and so we need to look at how the instruments described in the preceding two chapters are actually used around the world at such sites.

An important point to recognise from the start is that these traditional instruments were installed, in some cases over a century ago, for the purposes of weather forecasting and not for detecting small changes in climate over long periods; such thoughts were nowhere in sight. Even stations installed in the 1950s and 1960s were of that earlier generation and for those less demanding, traditional purposes.

These installations still suit their original objectives admirably, but they are much less well suited to the new role now demanded of them of detecting climate change. Since we have nothing else, however, we must do the best we can with them. What can be achieved will be looked at in Chapter 8, when we move on to see how the data from individual stations are checked and corrected (where possible) before being merged into areal means. In this chapter we look at how the instruments are operated in the field. First, a brief reminder of the instruments is useful before we look at how they are installed and managed.

The instruments

Thermometers

There are usually four thermometers in a screen (see Figure 3.4): a mercury-in-glass thermometer for *present temperature*, a similar thermometer with a wetted wick on its bulb – a *wet bulb thermometer* – to estimate relative humidity, a mercury-in-glass *maximum thermometer* and a spirit-in-glass *minimum thermometer*. These thermometers are (today) graduated in 0.5°C steps, with readings being estimated by eye to 0.1°C. The limitations and errors of these instruments were discussed in Chapter 2. In the UK prior to about 1970, the Fahrenheit scale was used and measurements were rounded to the nearest 1°F (0.56°C). Traditionally there is often also a bimetallic thermograph and a hair hygrograph in the screens, but these are of no use in climatology (Strangeways 1999, 2003).

Screens

As soon as the importance of shielding thermometers from radiation was realised, in the nineteenth century, a great variety of stands and screens were designed to house thermometers. Prior to about 1880 the accuracy of many temperature measurements is uncertain because we do not know what screens were being used, if any. Sometimes just north-facing walls (in the Northern Hemisphere) were used, which is inadequate because it is not just radiation from the Sun that must be avoided, but all of the long-wave IR radiation from everything in sight. While, after 1880, Stevenson screens were used in the UK, a look around the world's meteorological sites today will reveal a range of many different designs of wooden slatted screen, often quite different in size and height (see Figure 3.5). Nevertheless, by the 1880s, screens had become more or less standardised on some form of wooden slatted construction, whether it be to Stevenson's design or a variation on it. However, lacking good *metadata* (metadata are *data about data* – see Appendix E), we usually do not know what type of screen was in use at

Figure 4.1 In regions subject to snowfall, louvred screens become blocked and snow will also get inside the screen and cover the instruments. One solution is to fix doors to the sides of the screen, as illustrated, which can be closed on the windward sides when snow is falling. This of course reduces ventilation considerably. It also requires someone to be in attendance all the time.

most stations. If, however, we are looking at *changes* in temperature, rather than at absolute temperature (see Chapter 8 on the use of temperature anomalies) then, provided the screens were not changed, we might reasonably argue that it does not matter exactly what type of screen was used. Nevertheless it would have been much better if the screens had all been the same, and we should aim for this in the future (see Chapter 11).

Screens at tropical and polar sites

As noted in Chapter 3, John Aitken pointed out a century ago that Stevenson screens overheat in the tropics, with errors of up to several degrees being possible. Yet the majority of measurements from the tropics, used in deriving global climate datasets, originate from such screens. This is not a problem restricted to Stevenson's design: all naturally ventilated screens, including the miniature screens used with AWSs, suffer from the same defect. The problem also occurs at higher latitudes in strong sunlight and low wind, even in winter.

In regions that are subject to snow, all naturally ventilated screens are also rendered inoperative when snow blows into them or covers their outsides, ventilation being stopped. One preventive measure is to fit the screens with doors that

Figure 4.2 Under conditions of super-cooling in cloud, rime ice forms as soon as the droplets come into contact with any object. The ice can build up rapidly (into the wind) and can be very damaging to instruments. Under the rime in this photograph an AWS is completely encased. This prevents measurements being made in many mountainous areas, in this case on the Cairngorm Mountains in northern Scotland.

can be shut on the windward side (Figure 4.1), but this reduces the ventilation still further. Rime ice has the same incapacitating, and at times damaging, effect (Figure 4.2). It is very difficult to do anything about this (Strangeways 1985). While at manned sites, the snow and ice might (theoretically) be removed by hand (as at the old Ben Nevis summit station in Scotland), this cannot be kept up all day, every day. As with the tropical situation, it is not possible to say how much and in what way snow and ice might have affected the global datasets.

The problem of snow and rime affecting readings extends beyond temperature measurement; for example, wind speed readings are affected by the accumulation of snow and ice in and on anemometer cups (Strangeways 1985). The official line on this is that the cups can be cleared of snow by hand, but on tall masts this is rarely done and with automatic instruments it is clearly not possible. Many wind measurements in cold regions will be wrong (low) as a consequence. As to the measurement of snowfall itself, this is extremely difficult to achieve accurately; errors being much greater than for rainfall (Strangeways 2004).

Instrument calibration

To correct for the slow upward drift of mercury-in-glass thermometer readings due to the progressive contraction of their glass bulbs (brought about by changes in the molecular structure of the glass), as well as the slow changes to the chemistry of the liquid used in spirit-in-glass thermometers (see Chapter 2), it is necessary to recalibrate thermometers every few years. This should be done using a precision reference thermometer graduated in 0.1°C steps, in a mixture of water and melting ice. The reference thermometers must themselves be recalibrated, using water baths of accurately reproducible temperatures of which the triple point of water (0.01°C, or 273.16 K) is fundamental. The boiling point of water (equilibrium between the liquid and vapour phases of water) at a standard barometric pressure (of 1,013.25 hPa) is another precise reference temperature. There are others, as described under the heading *The International Practical Temperature Scale* in Chapter 2, but they lie well outside the range required for meteorological applications.

Having performed the calibration check, it is then necessary to keep a note of any changes and to remember to correct all subsequent readings. How many of the world's NMS thermometers have been recalibrated accurately and regularly and the necessary corrections applied thereafter is unknown, and a lack of metadata makes it impossible to check. The chances are that many will not have been – although one hopes that in developed countries they have – but this would need checking and cannot just be assumed. Thermometers in screens are fairly secure and are unlikely to have been broken and replaced, probably surviving for decades at established sites. The question of recalibration is thus not a trivial matter. How much the datasets may have been affected is impossible to say. The problem of sensor drift, whatever the type of sensor, needs to be avoided in future measurements (see Chapter 11).

All of the other instruments on a meteorological site also need recalibration; for example, a cup anemometer should be calibrated in a wind tunnel (to detect and correct for increased friction in the bearings due to wear and tear). The calibration of rain gauges is also necessary. Tipping bucket settings need checking, and for accurate measurements a different calibration factor is needed for different precipitation rates, the finite response time of the bucket becoming increasingly important as the rainfall rate increases. A one-factor-fits-all approach is not sufficient.

Instrument maintenance

All instruments on a meteorological site need periodic attention (additional to recalibration). They cannot be installed and left to their own devices thereafter, yet my experience is that many are. The maintenance of the instruments, and of the meteorological sites on which they are operated (Strangeways 2003), is steeped in the traditions, standards and attitudes of 50 or more years ago, and these do not meet the much stricter requirements for detecting small climate changes. Many sites have a very rough-and-ready feel to them, perhaps acceptable in simple meteorology but not for modern climatology. In addition, many countries have a slack record of maintaining anything, as a trip around the globe will confirm. I will not give specific examples of poorly maintained sites, since this would be undiplomatic, although not difficult. The *Ten Principles of Climate Monitoring* of the World Meteorological Organization's Global Climate Observing System (GCOS) is an attempt to bring stations up to a minimum standard, but its success is, reportedly, patchy (see Chapter 7). That the WMO considered it necessary to issue these guidelines is proof enough that sites need improving. But old habits are hard to change.

Of particular concern in the case of temperature is that of ensuring that the screens are maintained properly. The majority are 'naturally ventilated' by the wind, and this introduces errors when winds are low and radiation is high, the problem being compounded if the screens are not kept brilliant white by washing bi-annually and repainting on alternate years. Failure to observe this simple rule will cause the measured temperature to rise gradually year by year. This may not amount to much – perhaps a few tenths of a degree over a decade – and keeping the screens clean may seem an utterly trivial point hardly worthy of mention in comparison with the clever statistics used to process the data. But if the screens are not carefully maintained in this way, all the rest is undermined. All of our knowledge of global land temperatures comes directly, and solely, from these white boxes and their simple thermometers.

It should also be noted that the thermometer and screen errors just described are systematic, not random, resulting in an upward trend in the readings. When data from a number of stations are averaged (during the process of calculating areal means), the errors do not cancel each other out; all being upwards. In some cases these errors can be on a par with the observed climate-change signal. A technique used to detect errors is to compare data from neighbouring sites, but similar errors may be occurring at all of the surrounding stations. It is also difficult to know whether what appears to be an error is not in fact good data from a site with different characteristics. It is far better to be sure of the data in the first place, and while there is nothing we can do about the quality of past data, we can plan to collect better data in the future.

Met enclosures

The terms *meteorological site, met enclosure* and *weather station* are used interchangeably to describe the field-sites at which meteorological instruments are operated. While we are concerned here only with temperature, met enclosures also accommodate instruments to measure all of the other common meteorological variables, including precipitation, evaporation, wind speed and direction, sunshine duration (possibly also solar and net radiation) and ground temperature (Strangeways 1995, 2003).

Exposure of screens

There are certain key rules which need to be followed about how instruments should be exposed at a met enclosure. These rules are pretty much the same for all of the instruments, whatever they are measuring, and so the following remarks apply equally to thermometer screens, sunshine recorders, anemometers and rain gauges. The primary rule for all of them is that the instruments should be in a clear open space, without anything shading or shielding them from the sunlight and the wind; for example, a sunshine recorder must have a clear view of the sky to near the horizon in all directions where the Sun can appear. A rain gauge should not be closer to any object (tree, building, wall or another large instrument) than twice the height of the object, and preferably four times its height. An anemometer should be at a height of 10 m (higher if there are objects shielding it – the rules for this being complex; see Strangeways 2003), and if not at this height its readings must be corrected to what they would have been at the standard height.

The temperature screen should also be fully exposed all round so as to prevent a microclimate forming in its immediate vicinity and to ensure that the best ventilation possible is obtained for naturally ventilated screens. Sheltering by any objects will reduce ventilation and cause a rise in the measured temperature on days of high radiation. However, the top of a building is not appropriate, since temperatures change with height, especially at night, and the building itself will have an influence on the temperature. Nor is a hollow in the landscape or a steep slope suitable, as these may well have an untypical microclimate. Installation in a city or town can also bias the readings, and this bias may change as the city changes. In addition, readings in such places will not be representative of the countryside outside of the city limits (see the *Urban heat island* section below). It is traditional to operate screens over short-cut grass, but if the surrounding country is desert or rock or is snow-covered or is an expanse of water or dense forest, the screen should be deployed appropriately. The most important single rule is that measurement sites should be exposed to the weather all round and should not be shielded in any way.

These rules, however, are those specified by the WMO for conventional meteorological sites, used for meteorology and weather forecasting – where the aim is to get as accurate a measure as possible of the absolute temperature (wind, radiation, etc.), and a measure that is also representative of the surrounding area. For measurement of climate change, the requirement is different, for here it is *changes* in temperature (over a long period) that matter more than absolute values and, in that respect, a measure in a hollow, for example, would be acceptable, since it would be the change of temperature in the hollow that mattered. However, the hollow would probably not represent the surrounding countryside and so, in that sense, it would not be a good location. On a roof, similarly, the measure would not be representative and would also be liable to change if the building was extended or its use changed.

Representativeness of sites

Sites must first be selected to meet the local needs just described; that is, they need to be well exposed all round. But sites also need to be chosen that represent the countryside around them, for this is what is generally required. There may be exceptions where, for example, the conditions at a very specific site, such as beneath the canopy of a forest or in the middle of a town, are to be measured, but for meteorology – and especially for climatology – sites should represent the natural environment in which they are situated. This matter is examined more closely in Chapter 7 when considering the Global Climate Observing System (GCOS), but one matter concerning site representativeness needs more discussion here – the *urban heat island* (UHI).

The urban heat island and land-use changes

The question of urban heat islands and their effect on temperatures measured in urban areas is important and frequently misunderstood, so it is worth clarifying a few points.

Historically, NMS sites have been selected for convenience rather than for their representativeness. This is because a major influencing factor in choosing a site has always been the need to have an operator to attend the instruments at least once daily, and often throughout the day, and possibly at night also. This has resulted in most NMS sites being in or near towns. Additionally, NMS sites were installed largely for applications relevant to their time, such as the growing aviation industry, both civil and military, and for better weather forecasting. Placing stations at airports and in or near towns made good practical sense for all of these reasons and were appropriate to the needs of the mid twentieth century. At that time, precise climate measurements were not part of the agenda. Only since the climate (and climate change) became a concern in the 1970s has the precise location of weather stations become significant. Now that we are looking for very small and subtle changes in temperature over long time spans and over large areas, any systematic *changes* in bias introduced by the urban (or other atypical) environment becomes a concern. Not only is a temperature bias introduced by a city, no-one argues otherwise, but it is also likely that the city will grow with time, changing the microclimate around the instruments and thus changing the bias over time. It is important to stress this point. A fixed unchanging bias does not matter (in detecting climatic trends), but a changing bias brought about by the city changing does matter, since it introduces a trend that is not linked to climate.

The reason that built-up areas become warmer than the surrounding countryside during the day is because the incoming solar and terrestrial radiation (see Chapter 1) are divided into sensible and latent heat fluxes in different proportions from those in the natural countryside beyond the city. In large part this is because there is less water to evaporate in urban areas, the ground being covered with tarmac or concrete. There is also much less vegetation to transpire. This reduces the latent heat flux, with a higher proportion of the energy now going into the sensible and soil heat fluxes, that is into warming the atmosphere and buildings, resulting in higher daytime temperatures. The buildings and roads also store the day's heat more efficiently than vegetation and soil, releasing it slowly at night, thereby keeping the town warmer at night. If the city grows, these effects can increase, increasing the warm temperature bias.

However, much hinges on how 'rural' and 'urban' are defined. Peterson and Owen (2005) demonstrated that assessments of the rural/urban question were very dependent on the choice of metadata used to differentiate between the two. They concluded that to make such an assessment for all the weather stations used in the datasets would need suitable metadata from each station for the whole period since 1850, and such metadata do not exist. Jones et al. (1986) explained that, as it would be extremely laborious to look at the details of every station (even if such records existed), they have instead taken the approach of comparing adjacent stations in an attempt to detect serious mismatches that might be indicative of one station being in error. However, they comment that this is the cruder of the two possible approaches, but is the only realistic solution if thousands of stations are involved, with most lacking any metadata. This is undoubtedly the only realistic approach and is discussed more fully in Chapter 8.

In selecting sites suitable for the GCOS, 'rural' is defined by Peterson et al. (1997) as a settlement with a population of less than 10,000; a small town has 10,000–50,000 residents, while 'urban' is defined as having over 50,000. Despite this definition, a population of 10,000 does not, to my mind, suggest a rural settlement – it is a fairly large town. However, it is necessary to be careful, for while a station's name may be that of a large town, the station may not necessarily be in the actual town itself, just near to it.

A number of attempts have been made to detect and quantify the UHI effect. Four are outlined here:

1. One attempt to quantify the UHI effect involved comparing the difference in temperature trends on calm nights between urban and nearby rural sites, the logic being that on calm nights the UHI effect would be greater than on windy nights. There was indeed a difference between the two – there was an UHI effect. However, it was found that the difference did not change significantly with time, the long-term trend being the same at both rural and urban sites on calm nights (Parker 2004, 2006). The same tests were not done on daytime trends.
2. A comprehensive assessment of the effects of urbanisation on large-scale temperature trends in the USA was made by Karl et al. (1988). From 1,219 stations from the US Historical Climate Network (USHCN), sets of pairs of rural and urban stations were selected, and it was shown that there was an urban influence, with the bias being a non-linear function of population, with an urban warming of about 0.15°C over the period 1901–84.
3. Jones et al. (1990) assembled sets of data from the Soviet Union (38 stations), Australia (49 stations) and China (42 station pairs, making 84 in total). While the conclusion drawn from these case studies was that the influence of urbanisation was very small in those regions, Jones et al. suggested that the warming influences they had observed when recording local temperatures may not remain inconsequential in the future and that careful selection, inspection and monitoring for urbanisation will be necessary.
4. Very recently a study of the UHI effects in New York was undertaken by Gaffin et al. (2008). Their findings are complex, and not possible to summarise in a couple of sentences, but one of the conclusions was that the historical data reveal that the UHI effect is responsible for 33% of the overall warming that the city has experienced over the course of the twentieth century. It was also noted that in developing countries the effects are greater. This view is not shared by others and there would appear to be no simple and certain answer or agreement on the significance and implications of the UHI effect.

It has been stated by the IPCC (2007) that UHI effects are 'only localised in small areas' and that (therefore) they do not affect the overall climate. It is doubtless true that cities do not affect the overall climate significantly, if at all, but if the temperature is measured within an urban area, the readings could be biased by the immediate surroundings and this will be reflected in the areal mean for the whole area that the site is supposed to represent. The important point here, however, needs to be stressed once again – it is not the bias that is important; this is known to occur. What matters is whether the bias changes over time as the city grows and changes. The chances are that the bias probably will increase, since cities are growing ever more rapidly as the world's population grows, especially in developing countries. We are not living in an unchanging world as regards the number of people.

Perhaps the concluding comment by Jones et al. (1990) sums the situation up best:

> This concern [about the UHI] could be greatly lessened by an international effort to monitor and place observing stations outside of urban areas.

This was written in 1990, but their suggestions have not been taken up widely, if at all. This, of course, could only help improve future measurements, and the UHI continues to be a problem when using data from the past. Since the UHI may be a problem, it would be best not to use any stations sited in urban areas to produce datasets for the assessment of past climate change, but this is easier said than done.

Other changes in land use, such as in agricultural practice, crops grown and forest clearance, are also liable to affect temperatures measured within, or near, areas where they are happening, although this has been less well investigated.

Met enclosure maintenance

Short-term, local maintenance

As well as taking care of the instruments (as above), the site itself must not be neglected, and day-to-day attention to matters as simple as regularly cutting the grass without dislodging the rain gauges needs attention. A quick glance at a site reveals a lot. All too many sites show a lackadaisical approach to maintenance, reflecting the requirements of an earlier age, often looking more 'agricultural' than scientific. This suggests that less visible matters, such as instrument maintenance, might also have been neglected. Much depends on the personnel active at any one time and this can change as interest and commitment waxes and wanes. I have seen such changes at sites I have known for decades.

Long-term stability

Another important role that maintenance plays is in ensuring that the sites do not change over the years. Slow changes are not obvious, small bushes eventually becoming large trees, and since these do not happen overnight, they go unnoticed. Yet the slow growth of trees, and the gradual appearance of one building and then another, affect the local microclimate and so bias the measurements, the biases changing with time. These changes also act to reduce wind flow through the screens, and any of this can happen at 'rural' sites as much as in towns. Equally problematic can be conditions outside the enclosure, which are usually not under the control of the site operator.

To conclude this chapter, Figure 4.3 illustrates the very different style, size, condition and surroundings of just a few of the many met enclosures that I have worked on or visited in the UK and around the world. A recent article I wrote for *Weather*

(Strangeways 2009) describes how met enclosures can now be viewed on *Google Earth*, giving researchers a bird's-eye view of the GCOS (and many other) stations, enabling a judgement to be made about their microclimate and the surrounding countryside.

(a)

(b)

Figure 4.3 (a) A typical met enclosure on high moorland in Northumbria. This well-exposed site includes, in addition to the small Stevenson screen, wind, rain and sunshine instruments. (b) On farmland in southern England, this enclosure includes an automatic weather station (centre mast deploying wind, radiation and temperature sensors), along with the conventional screen, sunshine and solar sensors, soil temperature thermometers and rain gauges (out of view).

(c)

(d)

Figure 4.3 (*cont.*) (c) This Global Climate Observing System (GCOS) station is at the National Meteorological Service (NMS) at Port Moresby in Papua New Guinea. The international airport runway is just beyond the trees and in the distance are mountains covered in tropical rainforest. This site contains many instruments additional to the screen, including several raingauges, wind sensors and an evaporation pan. The city is close to hand, but the site could not be described as urban. (d) Near Tulkarm, West Bank, this site has a large screen, shown in close-up in Figure 3.5. The environment is semi-arid and rural, with olive groves and is well-exposed all round, although close to a small town.

(e)

(f)

Figure 4.3 (*cont.*) (e) Near Jaipur, India, the encroaching trees will reduce wind speed and aspiration of the screen. The monsoon rains were starting during this visit and the problem of keeping the ground in good condition was obvious. Unusually, the rain gauges were installed in concrete blocks. (f) In semi-arid conditions near Tripoli, Libya; blowing sand could be a problem. Note the height of the rain gauges and the large screen. (g) This enclosure is on the Tiéte River in Brazil at a hydroelectric dam site. This is a wide open site and is unlikely to change over time. The screen (*left*) is larger than the standard Stevenson screen.

(g)

Figure 4.3 (*cont.*)

References

Gaffin, S. R., Rosenzweig, C., Khanbilvardi, R., Parshall, L., Mahani, S., Glickman, H., Goldberg, R., Blake, R., Slosberg, R. B. and Hillel, D. (2008) Variations in New York city's urban heat island strength over time and space. *Theoretical and Applied Climatology* **94**(1–2), 1–11.

IPCC (2007) *Climate Change 2007: The Physical Science Basis.* Contribution of Working Group I to the Fourth Assessment Report of the Intergovernmental Panel on Climate Change. Cambridge University Press.

Jones, P. D., Raper, S. C. B., Bradley, R. S., Diaz, H. F., Kellyo, P. M. and Wigley, T. M. L. (1986) Northern Hemisphere surface air temperature variations: 1851–1984. *Journal of Climate and Applied Meteorology* **25**(2), 161–79.

Jones, P. D., Groisman, P. Ya., Coughlan, M., Plummer, N., Wang, W.-C. and Karl, T. R. (1990) Assessment of urbanization effects in time series of surface air temperature over land. *Nature* **347**, 169–72.

Jones, P. D., New, M., Parker, D. E., Martin, S. and Rigor, I. G. (1999) Surface air temperature and its changes over the past 150 years. *Reviews of Geophysics* **37**(2), 173–99.

Karl, T. R., Diaz, H. F. and Kukla, G. (1988) Urbanization: its detection and effect in the United States climate record. *Journal of Climate* **1**(11), 1099–123.

Parker, D. E. (2004) Climate: large-scale warming is not urban. *Nature* **432**, 290.

Parker, D. E. (2006) A demonstration that large-scale warming is not urban. *Journal of Climate* **19**, 2882–95.

Peterson, T. C. and Owen, T. W. (2005) Urban heat island assessment: metadata are important. *Journal of Climate* **18**(14), 2637–46.

Peterson, T. C., Daan, H. and Jones, P. (1997) Initial selection of a GCOS surface network. *Bulletin of the American Meteorological Society* **78**(10), 2145–52.

Strangeways, I. C. (1985) *The development of an automatic weather station for use in arctic conditions.* PhD thesis, Department of Meteorology, Reading University, UK.

Strangeways, I. C. (1995) Back to basics: the 'met enclosure'. Part 1. Its background. *Weather* **50**(6), 182–8.

Strangeways, I. C. (1999) Back to basics: the 'met enclosure'. Part 4. Temperature. *Weather* **54**(9), 262–9.

Strangeways, I. C. (2003) *Measuring the Natural Environment* (2nd edn). Cambridge University Press.

Strangeways, I. C. (2004) Improving precipitation measurement. *International Journal of Climatology* **24**, 1443–60.

Strangeways, I. C. (2009) Using *Google Earth* to evaluate GCOS weather station sites. *Weather* **64**, 4–8.

5

Measuring sea surface and marine air temperature

The entire heat capacity of the Earth's atmosphere is equivalent to that of the top 3 m of the ocean's water (Houghton 1997). And below the surface lies another 2,000–5,000 m, with trenches down to 11,000 m. In addition to their great depth and thermal mass, the oceans also cover 71% of the Earth's surface, the Pacific Ocean alone covering almost half of the planet. Given this dominance, measurements over the oceans are vitally important for climate evaluation, arguably more so than over land.

A brief history of measurements at sea

Meteorological and oceanographic observations have been made at sea since the seventeenth century (Kenworthy and Walker 1997), but it was not until the International Meteorological Conference in Brussels in 1853, largely at the instigation of Lieutenant Matthew Fontaine Maury of the US Navy, that weather observations from ships were put on a scientific footing. As a result of this conference, the British Meteorological Office was also established as a department of the Board of Trade in 1854, headed by Captain (later Vice-Admiral) Robert FitzRoy (Barlow 1994, 1997), previously Captain of the *Beagle*, which took Charles Darwin on his voyage of discovery. By 1855, FitzRoy had obtained the cooperation of 105 merchant ships and 32 ships of the Royal Navy in making weather observations at sea, equipping them with the necessary instruments.

Measurements were initially recorded in a weather register agreed at the Brussels conference. This was improved in 1874 by the Marine Superintendent of the Meteorological Office, Captain Henry Toynbee, who drew up the meteorological log, which was then recognised internationally. Observations were made six times a day, at the end of each 'watch'. These logs remained unchanged for about 45 years, but after 1921 they were gradually discontinued in favour of observations made every three hours (at 00:00, 03:00, 06:00, etc.). These observations were also transmitted by radio. In 1953, and again in 1982, the layout of the log was changed, resulting in today's meteorological logbook, which combines observations made and radio messages sent. Small changes were made in 1988, 1990 and 1994. Because of the importance of maritime meteorology, the International Meteorological Organization was set up in 1872, this later becoming the World Meteorological Organization (WMO) in 1951. Air temperature was recorded on the ships as well as the temperature of the sea surface, in addition to barometric pressure, using a mercury barometer developed by FitzRoy (Strangeways 2003).

Air versus sea temperatures

Over the land, air temperature is measured at a height of 1.5 m by thermometers inside Stevenson-like screens, the mean daily temperature usually being the average of the maximum and minimum readings, although there are variations on this. For compatibility, it might appear best for climate analysis to use the temperature of the air over the sea – the marine air temperature (MAT) – since it could be combined directly with land air temperatures (LAT) (Jones 2008). However, as shown below, it is difficult to measure air temperature over the sea, and so the sea surface temperature (SST) is generally used instead, since it is slightly easier (although still not easy) to measure. A definition of SST is given below. In some ways SST is better than air temperature because it changes more slowly and is less spatially variable, reducing the number of measurements needed. Both arguments have their strengths and weaknesses. While air temperatures *are* measured over the sea both from ships and from buoys, the bulk of the data from the oceans are measurements of the temperature of the top metre or so of the sea.

The use of SST as a proxy for MAT requires that there be substantial agreement between the two, despite their very different natures. Many tests have shown that MAT anomalies agree well with SST anomalies, but only on seasonal and longer time scales and only in the open ocean (Rayner et al. 2003). Even in the open ocean some regional discrepancies in trends have recently been found in the tropics (Christy et al. 2001). So there is still uncertainty about the strength of the link between SST and MAT, in no small part because we have few good air temperature measurements to compare with the SST.

The relationship between air temperature and the temperature of the surface of the sea has some similarity to that between air temperature and that of the land. There is often a difference of several degrees between air temperature at screen height and the surface temperature of the land, and so it is over the sea. When we talk of surface temperature, it is important to note the difference between the temperature of the 'skin' (outer few millimetres) and the bulk of the land or sea beneath the skin (see next section). This is particularly relevant when attempting to measure air temperature from satellites, since satellites only measure the very outer skin temperature of the sea or land surfaces and it is not a simple matter to derive air temperature from them, complex algorithms being necessary (see Appendix D).

The reasons for the differences in temperature are obvious. Recall, from Chapter 1, that the land and sea are warmed by the incoming solar and terrestrial IR radiation, the resultant heat from the warmed surfaces being transferred to the air by molecular diffusion and then by turbulence and convection. But these heat transfers lag behind the heat input, resulting in the surface being warmer than the air during the day. At night the surface is likely to cool, through IR radiation, and this will then cool the air in contact with it, although the surface will tend to remain colder than the air. In addition, the air may have arrived swiftly from a distance so it may be warmer or cooler than the land or sea that it passes over and will thus warm or cool the surface, and in so doing it will cool or warm itself. A further major factor is the evaporation of water (the latent heat flux); this complex interchange of energy at the surface is the subject of many research projects, for example those of Kent and Taylor (1995) and Yu and Weller (2007), the latter finding that there has been a long-term increase in the latent heat flux over the oceans of about 10% between 1981 and 2002.

To understand and quantify this interchange of energy at the surface requires sustained, systematic and consistent measurements, yet the available data concerning such energy exchanges over the oceans are not of the required quality and quantity to give reliable results, although improvements are ongoing (Zhang et al. 2007), including the development of radiometers for use on ships (Donlon et al. 2008).

Definition of 'sea surface temperature'

The top 10 m of the ocean has a complex and variable temperature structure affected by waves and air–sea fluxes of heat, moisture and momentum. The definitions of SST developed by the Global Ocean Data Assimilation Experiment High Resolution Sea Surface Temperature Pilot Project (GHRSST-PP) are as follows (Donlon et al. 2007).

The skin temperature

The skin temperature (SST*skin*) is the temperature measured by an infrared radiometer at wavelengths of 3.7–12.0 μm, and represents the temperature within the conductive diffusion-dominated sub-layer at a depth of about 10–20 μm (recall the laminar boundary layer around land surfaces). This temperature is subject to a large diurnal cycle.

The sub-skin sea surface temperature

The sub-skin temperature (SST*subskin*) is the temperature just below the conductive laminar sub-layer. This can be measured approximately using a microwave radiometer in the 6–11 GHz frequency range, but the relationship is neither simple nor constant.

The surface temperature at depth

All measurements of temperature beneath the SST*subskin* are known as *depth temperatures* (SST*depth*). These can be measured with buckets, from engine-room intakes, with hull sensors and from buoys (see below). They are distinct from those measured using remote-sensing techniques and must be qualified by a measurement depth in metres.

The foundation temperature

The foundation temperature (SST*fnd*) is the temperature of the water column that is free of diurnal temperature change. It is the temperature from which the growth of the diurnal thermocline develops each day, although on some occasions there is no clear SST*fnd*. Only *in situ* techniques are able to measure this temperature. Algorithms have to be used to derive the SST*fnd* from remote-sensing instruments; they cannot measure it directly (see Appendix D). SST*fnd* is simply a more precise definition of what is normally referred to as the 'bulk temperature'; that is, the temperature of the water not affected by transient surface variations.

Ships as instrument platforms

Ocean weather ships (OWS)

For a few decades in the twentieth century, dedicated ocean weather ships (OWS) were operated, the sole purpose of which was to observe and report the weather from fixed locations. For a time there were ten ships in the Atlantic and three in the Pacific, but by 1974 the US part of the operation was reduced to three Atlantic and one Pacific station, and by 1976 all US ships were discontinued, some being replaced by automatic buoys. European and Canadian ships continued for a little longer, but they too had been discontinued by 1981.

Norway, however, has continued to operate 'Station M' from the ship *Polarfront* at 66°N 2°E in the Norwegian Sea. It has collected much useful and unique data on many variables, including the temperature and salinity of the water at several depths (Dickson and Østerhus 2007).

The voluntary observing fleet (VOF)

Today, ships are used both for specialised research, when the ship is dedicated entirely to that purpose, and for routine measurements for meteorology and weather forecasting. All these routine observations, from merchant ships, are made on a voluntary basis, so as to ensure the required dedication and high quality required. The ships are known as voluntary observing ships (VOS), jointly as the voluntary observing fleet (VOF), and the observers as voluntary marine observers (VMOs). Each National Meteorological Service (NMS) recruits its own ships, and there are regular visits to them by foreign NMS personnel to supply materials and attend to instruments. It is a pity there is not this same exchange with regard to land stations.

There are three classes of ship in the VOF: *Selected ships* observe wind, 'weather' (a summary of present and past weather in a specially coded form), barometric pressure (and its tendency), air and sea temperature, humidity, clouds and waves. Certified instruments are used. *Supplementary ships* make the same observations but collect less data about pressure tendency, sea temperature and waves. *Auxiliary ships* are similar to supplementary ships but do not report cloud and do not have certified instruments. The *Marine Observer's Handbook* (Met Office 1995) describes all these in detail, as well as the observation of sea-ice and other marine and atmospheric phenomena. Of these observations, only barometric pressure, air temperature, humidity and sea surface temperature are measured by instruments, with the rest being visual observations. Routine meteorological reports from merchant ships are collated in the International Comprehensive Ocean–Atmosphere Data Set (ICOADS) by NOAA's National Climatic Data Center (NCDC) (Woodruff et al. 1998; Worley et al. 2005).

Decline in the number of voluntary observing ships

Like the OWS, VOS have also reduced in number, with a peak in 1984/85 when there were about 7,700 ships worldwide. Since then there has been a decline, and in June 1994 the VOF had dropped to about 7,200 ships. Numbers have continued to fall and currently it is estimated that there are only about 4,000 observing ships worldwide. Furthermore, reports from the VOS are heavily concentrated along the major shipping routes, primarily in the North Atlantic and North Pacific Oceans, with data-sparse areas in all of the Southern-Hemisphere oceans. Today more ocean data come from various types of buoy, which are described later, rather than from ships.

The reasons for the decline in VOS numbers is a combination of factors, in particular a greater reliance by the weather forecasting community on buoys and satellites. VOS operators have also become concerned that their data are appearing on the Internet in a form that allows them to be identified, along with their positions. This raises concerns about both commercial confidentiality and terrorism, and the number of attacks by pirates is also on the increase. Ship owners point out that they were asked to provide reports for weather forecasting and climate research, not to have these appear on the Internet. But VOS reports are classified as essential data, and so they must be freely exchanged. Kent and Taylor (2006a) are looking into how the decline in VOS can be dealt with by addressing these problems.

There is also a considerable variability in the quality of data from VOS and they are seen as variable, unreliable and expensive. Also, to cut costs, the number of Port Meteorological Officers (PMOs) is rapidly decreasing, although the VOS system relies on them. There is also an increasing trend to rely on ships with automatic weather stations (AWSs), hourly observations from these in real time being preferred by weather forecasters. But there are fewer ships using AWSs than making manual observations, resulting in a reliance on only a subset of the VOS. This also results in a loss of non-instrumental data such as cloud observations, which are not yet made automatically. Nor are the AWSs being introduced in accordance with the Global Climate Observing System's (GCOS) *Ten Monitoring Principles* (see Chapter 7). Hence even the ships with AWSs have their problems. This lack of standardisation and consistency will be raised again in the final chapter. The VOS Climate Project (VOSClim) aims to establish a high-quality climate dataset, along with details of each ship's instruments.

Measuring air temperature on ships

Marine air temperature (MAT) (and humidity) is measured in the same way on a ship as on land, but using a miniature Stevenson screen with (single-) louvred walls, just large enough to hold the wet and dry thermometers and nothing more (see Figure 3.7). No chart recorders or maximum and minimum thermometers are used, spot observations being made at set times. The screen has to be exposed as far as possible away from any heat sources, the rails on the bridge being a typical choice, ideally with a screen on each side of the ship to allow the windward side to be used, although this duplication is seemingly rare. There is considerable variability, however, as to where the screens are exposed, some very inappropriately near doorways and hatches. Nor can there be the same consistency of exposure as is possible on land, since ships differ in size and structure, with many changes occurring in their design and height over the years. On large modern vessels, the temperature is measured tens of metres above the sea surface and this introduces some uncertainty. (As was noted with regard to Stevenson screens on land, height is important because of changes in temperature with elevation.) Kent and Taylor (1991), in a report for the National Oceanography Centre, Southampton (NOCS, formerly the

Southampton Oceanography Centre (SOC)), provide a catalogue of ships – their routes, type, size, speed and country of origin – along with drawings of ships and details of the instruments used, including the height of screen. Adjustments for height can be made using boundary-layer theory (Rayner et al. 2003).

In addition, the ship will generally be warmer than the surroundings due to heating by the Sun by day and because of its internal power generation at all times; there is, in fact, a 'heat island effect' equivalent to that of cities and towns (see Chapter 4). Because of these problems, air temperature measured on ships is not considered reliable enough to use as a direct counterpart to air temperature measurements over the land, although Berry et al. (2004) and Berry and Kent (2005) are developing techniques to rehabilitate these data. Again we see, however, just how much adjustment has to be made to the raw data to make them usable, further strengthening the case for much better arrangements in the future. Furthermore, in the nineteenth century, most thermometers on board ships were not screened at all and so there is doubt as to the reliability of readings from the days of sail (Chenoweth 1996).

Temperatures measured from ships during the hours of darkness – night marine air temperatures (NMATs) – are considered to be more reliable because there is no heating effect from the Sun on the ship, and so they are used more widely than MATs. Rayner et al. (2003) and others use the period from one hour after sunset to one hour after sunrise to minimise the influence of deck heating, which lags behind the solar input. But this is not particularly satisfactory since half the day is lost and a daily mean is consequently not available (Jones 2008). Also, all days when there is little solar radiation or days when there is a strong wind, which could have yielded day-long measurements, are also rejected when readings might be no more prone to error than those on the land under similar conditions (except for their height). Even though free from warming by the Sun, the ship itself is a source of heat at all times and this could affect even the night-time readings.

Measuring sea surface temperature (SST) from ships

Bulk and foundation SST is the temperature of the top few metres of the sea, as defined earlier. It is measured from ships in three different ways: (1) a water sample is collected in a bucket and measured on deck with a thermometer, (2) the temperature of the engine-room intake (ERI) cooling water is measured manually or automatically, or (3) the temperature of the ship's hull below the water line is measured automatically. Up until the 1940s, buckets were the most common method, but since then the use of ERI temperatures has increased, becoming more numerous than buckets by the early 1980s, although buckets are still in wide use. ERI temperature was, however, measured on a few American ships as early as the 1920s. Measuring hull temperature is much less common (Kent and Taylor 2006b).

Recently, shipboard radiometers have been developed to measure the infrared emissions from the sea surface, in part to validate SST measurements made from satellites (see Appendix D). But the measurement is not direct and requires an algorithm to convert to skin SST and thence to bulk temperature. The raw radiometer measurements are affected by changing atmospheric humidity, rapid warming by solar radiation, sensor drift, the angle of deployment and variations in seawater emissivity (Donlon et al. 2008). As is the case with all remote sensing, it is a long way from the basic measurement to a quantification of the variable itself. A combination of remote sensing with ground-truth gives the best results.

Buckets

In this rough-and-ready method, a bucket is thrown into the sea on a rope and hauled back onto the deck when full, whereupon a thermometer is stirred around in it for a few minutes. A mercury-in-glass thermometer, of a similar type to that used to measure air temperature in a Stevenson screen, is suitable for the purpose, although it may have some protection around it to prevent damage.

Before the Brussels Maritime Conference of 1853 agreed on the standardisation of observations at sea (Maury 1858), usable data are scarce. For instance, in 1821 Otto von Kotzebue (1787–1846) described the measurement of SST as follows: 'The temperature of the water at the surface is easily found, by a common thermometer let down a few feet, and immediately drawn up again.' Chenoweth (1996) comments, in his article on nineteenth-century marine temperature data, 'It is impossible to determine how common this technique was, but it does suggest our imperfect knowledge of actual practices. No bucket appears to have been used.'

But the agreed 1853 standards, unfortunately, allowed a certain amount of leeway, and this led to different countries using different buckets (unfortunately, this sort of thing is still going on today with AWSs). New standards also evolved with time, each country adopting its own modified techniques, the result being a range of different water collection methods.

Initially, buckets were simple wooden constructions with the appearance of half-barrels on a rope, but these were gradually replaced by lighter canvas buckets. Some of these had an internal metal cylinder to provide rigidity and so allow more effective filling with water. Later a double-skinned canvas construction was introduced, the outer layer insulating the contents from cooling due to evaporation and to heat loss to cooler air, and from warming due to heat gain from warmer air and from direct or reflected solar radiation. Tests, however, have shown that the insulation is not perfect (Kent and Taylor 2006b). Figure 5.1 illustrates three of the buckets now in use.

Figure 5.1 Selection of buckets: UK Met Office insulated bucket (*left*), German metal and leather bucket (*centre*), UK Met Office MK II canvas bucket (*right*) (reproduced by permission of the UK Met Office).

As the speed of ships and their height increased, the use of lightweight canvas buckets become increasingly difficult, the buckets being liable to skim along the surface, resulting in the very top layer, or just sea spray, being collected, neither of which will be at a representative temperature. The UK Met Office now supplies a heavier rubber model to ensure that the bucket submerges 'cleanly' to a depth of about a metre.

The best method of collecting water is to throw the bucket so that it lands towards the bow of the ship, to let it sink to the required depth, and to pull it out as it passes beneath, while still travelling at zero speed relative to the water. Once the bucket is dragging in the water, it is difficult to resist the pressure of the water moving past it. This makes it more difficult to allow a period of bucket temperature stabilisation in the water before removal. It is also necessary to avoid contamination of the sample with water from the ship's engine-cooling water outlet at the stern.

There are large differences between SST measured by these different buckets, most of the errors arising from the exchange of heat between the water, the bucket and its surroundings. This varies from bucket to bucket and with the procedures adopted. Take, for example, the initial state of the bucket. It will most probably be warmer or colder than the sea into which it is about to be lowered. It may have stood in the sunlight, been in a warm room, been wetted by rain and cooled by evaporation, or have stood in a wind which is colder or warmer than the sea. When water enters the bucket it will be either warmed or cooled as the temperatures equalise, introducing an error. This type of error can be lessened by allowing the bucket to take up the same temperature as the sea by taking two successive samples, one after the other, with a stabilising period between, but this appears to be rarely done.

These errors will tend to be dependent on location and season, making them more difficult to correct. As is the case with land stations, metadata are not universally available, although the WMO have compiled a catalogue of observing ships (Kent and Taylor 1991) and this has allowed more corrections to be made than would otherwise have been possible. It is worth reflecting on the fact that the vast majority of work being done to investigate climate change is directed at sorting out the considerable confusion that arises from the numerous errors that each instrument and technique introduces, and then applying a correction factor. The way to avoid this in the future is dealt with in Chapter 11.

Much work has been done to assess the accuracy of SST measurements made with buckets and to apply corrections. Folland and Parker (1995) developed an empirical method of correcting SST measurements based on models of heat and moisture transfers from canvas and wooden buckets exposed on deck. One of the models was tested using actual measurements on board ship and in a wind tunnel. This work resulted in geographically and seasonally varying corrections for the period 1856–1941. The global annual mean bias correction increases steadily, from 0.11°C in 1856 through 0.17°C in 1872 to 0.30°C around 1905 and 0.39°C by 1920, remaining at around 0.40°C until 1941. Folland (2005) used a climate model to show that land air temperature trends over the continents were much better simulated with bucket corrections than with uncorrected SSTs. These corrections were subsequently slightly revised by Rayner et al. (2006) following the incorporation of additional data with a different mix of observing methods into ICOADS (see above). Further corrections applicable to the period from 1945 to about 1960 are also needed (Thompson et al. 2008). Yet again, one is struck by the large amount of work that has to be undertaken in an attempt to correct past readings. The future must be better planned.

Parker et al. (1995) described the varying coverage of SST and MAT measurements from the mid nineteenth century to 1995 and reviewed the methods used to remove systematic biases. In the same paper, they present global datasets of SST and night-time MAT anomalies for 1856–1994 using the best bias-corrections available, and assess the sensitivity of the datasets to imperfect coverage. They conclude that their results demonstrate the need for improved methods to extract the signal from incomplete data.

There have been, and still are, so many different procedures, types of bucket and sizes of ship, all operating in a wide range of climates, that complete correction of the data would seem to be an unattainable goal. It strengthens the argument that future measurements must all be made in the same way and with identical equipment everywhere. While it is recognised that such an aim may be thwarted by commercial considerations and human and national pride, it must be the aim. How it might be achieved is discussed in the final chapter.

Engine-room intakes

The temperature of the water pumped in to cool the engines – the engine-room intake (ERI) temperature – was being used as an alternative to buckets even in the early twentieth century by some ships in the US navy. During the Second World War, ERI measurement was preferred because of the danger of using lights at night but, since then, buckets (mostly insulated) have continued to be used and are still in use today, although the number of ships measuring the ERI temperature has gradually increased, their number exceeding those using buckets by the early 1980s. The ERI temperature can be measured manually, with a mercury-in-glass thermometer, or automatically with a platinum resistance thermometer (PRT; see Chapter 2).

The exact location of the measurement point is important, because if it is too near to the engines (after having travelled along pipes inside the ship), warming of the water will result. There is also a tendency for observers to round off manual readings, sometimes to the nearest degree (apparently more so than with bucket readings; see Kent and Taylor 2006b) and parallax errors in reading the mercury-in-glass thermometers have also been reported (which could also affect bucket readings). Increasingly a PRT is used to measure the temperature automatically. As the ship rolls and pitches, water is drawn from different depths, and this leads to some error as there is often a layer of water near to the surface which becomes warmed during the day.

Hull temperatures

A PRT fixed to the hull, a metre or so below the normal water line, can be used to measure SST, the conduction of the steel being good enough for it to take up the temperature of the outside water. As the hull has a large mass, the mean temperature is sensed and this compensates for any rolling or pitching of the ship. This technique has only come into use since the 1940s, and is still a rarity, despite being the best of the methods available. Due to the infrequent use of the method, little work has been possible to assess its accuracy and to compare it with buckets and ERI. However, Kent et al. (1993) found that SSTs from ERI were biased on the warm side relative to hull temperatures by 0.35°C on average.

Differences between bucket, ERI and hull measurements

A comparison of the methods is not simple because metadata are not always available and, when they are, they may be incorrect. There is also the complication that the method used to make the measurements may vary on the same ship, two methods being in regular use; for example, while buckets are the normal method, in bad weather or when other duties take precedence, ERI may be used instead, with no record as to which was used.

There are significant differences between SST derived from buckets and ERI, with a strong seasonal cycle between the two (Roll 1951; Folland et al. 1993; Kent et al. 1993), although Kent and Taylor (2006b) showed that the difference between the two methods reduced over the period 1970–98, probably due to improvements in ERI measurements. Prior to 1990, the mean differences between buckets and ERI over the North Atlantic varied between 0.5°C and 0.1°C, the buckets being the cooler. However, it is possible that the engine intakes are biased relatively cool in summer and warm in winter, as they measure deeper water, probably below any daily shallow thermocline. Since 1990, the differences are much reduced, probably due to better positioning of the ERI thermometers, along with the introduction of remote readouts or automatic recording.

It is clear from the extensive literature on the subject that uncertainty remains about the relative accuracies of the various methods and between the various buckets used. The uncertainty also changes over time from the 1850s onwards and with the season, with the particular ocean in question and with which country made the measurements; how much better it would have been if one method had been used throughout. But, as with measurements over the land, planners did not have climate measurements in mind when the measurements started to be made and such small matters were irrelevant. Again we are being forced to use data that were never intended for the precise task of detecting small changes over long time spans. The National Oceanography Centre, Southampton (NOCS) express the problem thus:

> Climate researchers today rely more and more on data that have been collected operationally for a variety of applications including for weather forecasting. Although the adequacy of the observations for the immediate purpose is already assessed routinely, the adequacy for applications downstream, such as the development of climate datasets, is not currently assessed as part of the operational system (NOCS 2009).

A submersible electrical thermometer

Lowering a PRT into the sea on a cable would be no more difficult than lowering a bucket, indeed it would be easier. It would have been possible since 1970, or earlier, to design an aerodynamic housing that made the sensor submerge neatly at a controllable depth, the temperature being read on a hand-held display to an accurately calibrated 0.1°C. It is surprising that such a simple solution has not been developed and adopted internationally; it would cost little and would be simple to use. An *in situ* measurement is an obvious answer to the difficulties experienced with buckets and ERIs, there being no errors due to warming or cooling of the samples. While ship measurements continue to be made, it would be useful to adopt such a new instrument to obtain accurate measurements; it could also be designed to measure air temperature a few metres above the sea surface (with the sensor protected from spray and rain).

That buckets are still being used, along with ERI and a few hull temperature measurements, is an example of the conservatism of the meteorological community and the apparent inability of its participants to cooperate globally and agree on common methods. This bodes ill for the future. And to these established methods have now been added buoys, with their different errors, further complicating the situation.

Buoys as instrument platforms

The majority of SST measurements are now made by buoys of one type or another, including moored deep-sea buoys, moored inshore shallow-water buoys and by various designs of drifting buoys, the latter now being the most common. We look at each of them in turn.

Moored buoys

The UK situation will be used as a basis for a general consideration of this genre of buoys. The Met Office operates two types of moored buoy, those in deep water and smaller buoys used around the coasts (Met Office 1997).

There are 11 moored buoys operating in the open ocean in water depths from 2,000 to 3,500 m, with two being operated jointly with France in the Bay of Biscay. There are two smaller inshore buoys in the Irish Sea along with five lightships in the English Channel. Three stations are also installed offshore of small islands to the north of Scotland. The USA and Canada operate similar buoys off their Atlantic and Pacific coasts and across the Pacific, America currently having about 65.

Figure 5.2 illustrates a deep-water moored buoy to the UK design, being visited by the *Salmaid*, a ship used by the Met Office to install and service this type of station. The hull of the buoy is 3 m in diameter with a sensor ring 4.5 m above sea level (Met Office 1997). The variables measured are barometric pressure, air

Figure 5.2 Deep-water moored buoy operated by the UK Met Office on the edge of the European continental shelf (reproduced by permission of the UK Met Office).

temperature and humidity, sea surface temperature, wind speed and direction, and wave height and period. With the exception of the wave sensor, all sensors are duplicated in case there are failures, as service visits are expensive and limited to a set schedule.

At present there are four shallow-water moored buoys in the Irish Sea and the English Channel (Figure 5.3). The sensors are the same as those on deep-sea buoys, but since the buoys are more easily visited, there is no need to duplicate the sensors. Use is also made of oil rigs, and around the UK coasts there are at present seven such installations, to the east and north.

One advantage that moored buoys have over ships (and over drifting buoys – see below) is that they are making measurements at a fixed point, which is an advantage over the ever-changing location of moving platforms. They also measure air temperature as well as SST.

Figure 5.3 Coastal-water moored buoy (reproduced by permission of the UK Met Office).

Air temperature measurements on moored buoys

The thermometers (PRTs) are housed in miniature screens of the AWS type, as Figure 5.3 shows, and measure air temperature to an accuracy of ±0.2°C. Although the seas are less sheltered from the wind than the land can be, and thus the screens will be better ventilated, calm conditions do exist, and at these times the screens will suffer the same heating problems as those on land (see Chapters 3 and 4). Ventilation will be assisted, however, by the swaying of the station due to wave motion. Under some conditions the evaporation of sea spray could cool the thermometer screens. The accompanying measurements of wave height and wind speed can allow adjustments to be made to temperature readings under calm and windless conditions. The humidity sensor is housed in a separate but identical screen to allow its exchange, during maintenance, independently of the temperature sensors. Unlike most land-based AWSs, the buoys use an ion-exchange humidity sensor. All the sensors, except the sea surface temperature and wave sensors, are changed every six months.

Sea surface temperature measured by moored buoys

A PRT sensor, fixed to the hull of the buoy, measures the water temperature at 1 m depth to an accuracy of ±0.2°C. Because the sensor is located at precisely the required depth, the data are truly representative of temperatures at 1 m. The swaying of the buoy will not affect the measurements because the hull has a large thermal mass that averages out the readings, as with hull sensors on ships. Such buoys, therefore, provide some of the best SST data available.

As an aside, it is worth noting that because precipitation is not required for weather forecasting, it is not measured on these buoys (although some buoys operated by the USA in the equatorial Pacific, for El Niño monitoring, do have rain gauges). Such measurements would be useful for climate research. Little is known about precipitation over the oceans.

Telemetering the measurements to base

Because the buoys are operated for weather forecasting, their measurements are telemetered in real- or near-real-time to a remote base. The moored buoys on the edge of the UK continental shelf use Meteosat to relay data to shore (EUMETSAT 1995). Like the sensors, the telemetry equipment is also duplicated to guard against failure, with each data collection platform (DCP) transmitting the measurements from both sets of sensors. Upon reception, the two sets of data are combined to give optimum results, which are then coded in SYNOP (surface synoptic observations) for transmission on the Global Telecommunications System (GTS) (a network operated by WMO for the interchange of meteorological data; see Chapter 7). On coastal buoys, telemetry can be by UHF line-of-sight terrestrial radio, or via satellite if out of direct radio range. Lead–acid gel batteries charged by solar panels power all the buoys.

Buoy location systems

Buoys moored in deep water do not move more than a mile from their point of fixing and so need no method of location. However, if one should break free and drift, the means to locate and retrieve it is necessary, owing to the high cost of replacement. Two methods are used, the first being the inclusion of a Global Positioning System (GPS) receiver as one of the channels transmitted by the DCPs. Alternatively an

Argos DCP (as opposed to a Meteosat DCP) can be used to transmit the data. The Argos system is a network of polar orbiting satellites in low Earth orbit which, by virtue of their movement relative to the ground, can locate the position of the DCPs (Strangeways 2003). Two radar reflectors are also fitted, allowing the buoy to be detected by local shipping for safety reasons.

Drifting buoys

Joseph Louis Lagrange and Leonhard Euler

In oceanography (and in fluid dynamics generally), the measurement of currents can be made in two frames of reference – Lagrangian and Eulerian. In the first, the instrument moves within the fluid and is tracked, an example being a radiosonde moving with the wind or, in the case under consideration here, a free-floating buoy. (The comparison with a radiosonde is not quite equivalent since the balloon has lift, which pulls the sensors upwards through the atmosphere; the comparison is purely in the sense that the balloon is carried along by the wind.) These are known as *Lagrangian* measurements.

An alternative is to make the measurements at fixed sites, with the fluids flowing past the instrument; for example, measurements from an anchored buoy. These types of measurement are known as *Eulerian*. These different concepts are based on the work of the two mathematicians Joseph Louis Lagrange and Leonhard Euler in the eighteenth century in connection with their general study of the flow of particles. The concepts are even used in the quantum world, to follow the flow of individual electrons and photons, and up into the realm of the orbits of planets. Since it would be difficult to install moored current meters across the oceans, the drifting, Lagrangian, option has been used to study ocean surface currents.

In the past, ocean surface currents were estimated by comparing the actual position of a ship, fixed by astronomical observations and timing (or today by GPS), with the expected position estimated by dead-reckoning (that is by measurements of speed, direction and transit time with correction for leeway-drift due to winds). The vector difference of the two is that caused by currents. This method is assumed to give an estimate of the current at about half the draught of the ship. It cannot be used over distances greater than about 600 km or over a time greater than a day, as several different currents may be involved. It is not accurate and is limited to shipping lanes. But, recently, drifting buoys have greatly improved our knowledge of currents, and with them also the SST.

Lagrangian buoys

In place of ships, the world's ocean currents can now be measured with drifting (Lagrangian) buoys. One of the earliest was made in 1975 as part of the North

Pacific Experiment (NORPEX). It was large, being 3 m long and 38 cm in diameter, and was made of fibreglass with a 9 m 'parachute' acting as the drogue at a depth of 30 m. In 1982 the World Climate Research Programme (WCRP) decided that a standardised, low-cost, lightweight, easily deployed surface drifter should be developed, this taking place under the Surface Velocity Program (SVP) of the Tropical Ocean Global Atmosphere (TOGA) experiment and the World Ocean Circulation Experiment (WOCE). By 1993 an agreed design for the SVP drifter had emerged, this becoming the foundation for future SVP drifter development (Sybrandy and Niiler 1991). Today the array of SVP drifters is known collectively as the Global Drifter Program (GDP), part of the Global Ocean Observing System (GOOS) and the Global Climate Observing System (GCOS) of the World Meteorological Organization and a scientific project of the Data Buoy Cooperation Panel (DBCP) of the Intergovernmental Oceanographic Commission (IOC). The scientific objectives of the GDP are to provide operational, near-real-time surface velocity, SST and sea-level barometric pressure observations. Present manufacturers of SVP drifters include Clearwater Instrumentation, Marlin-Yug, Metocean Data Systems, Pacific Gyre and Technocean.

The use of such buoys has increased considerably over the last decade and the greatest proportion of the SST data collected today is gathered by these drifting buoys, the number of observations now exceeding those made by ships.

Construction of the buoys

All of the drifting buoys are constructed as spherical floats of up to a metre in diameter, made of ABS (acrylonitrile butadiene styrene) plastic and stainless steel, below which is suspended a 'parachute' that acts as a drogue, dragging the float along with the current, the drogue being adjustable to depths of 5–100 m (typically 15 m) (Figure 5.4). What is descriptively known as a 'holey sock' drogue is used, these taking the form of cylindrical constructions up to 5 m long (deep) with holes on opposite sides which pull the float along with the current they are embedded in. The drogue has a large surface area that makes it hard for wind and waves acting on the surface sphere to pull it through the water. A drogue without holes is a cylindrical hydrofoil and it is possible for that shape to develop 'lift' from currents passing around it, which could pull the drifter through the water differently from the flow of the current. The holes in the drogue allow pressures that could produce lift to equalise through the drogue. A properly sized holey sock drogue can minimise the difference between true currents and currents estimated by a drifter to 2 cm s^{-1} (W.G. Williams, Clearwater Instruments, personal communication, 2008).

The SST sensor is normally a linearised thermistor, installed towards the bottom of the float by means of a through-hull fitting, measuring to ±0.1°C. There are some differences in buoy sizes, but the measured temperature is that of the water

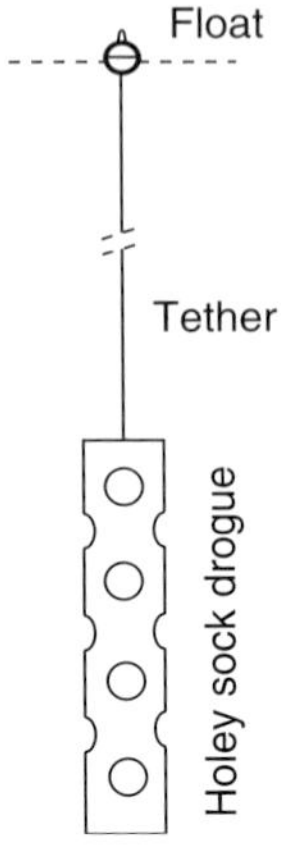

Figure 5.4 Drifting buoys are up to 1 m in diameter, attached to drogues up to 5 m long beneath, to follow the current. The depth of the drogue is typically 15 m.

between 20 and 30 cm beneath the surface. This is slightly less than the depth at which buckets measure the temperature, and much less than that of the ERI and hull temperatures, putting them at a depth which can be affected by diurnal swings of temperature. As measurements are made every hour, however, allowances can be made for the daily cycle and much useful additional information gleaned about the cycle. Some of the drifting buoys can also be equipped with an array of thermistors to measure temperature profiles down to greater depths. These can be used to establish the diurnal cycles of SST, which can in turn be used to correct the readings of other drifting buoys and of satellite observations (Kennedy et al. 2007).

As with the moored buoys, drifters must transmit their measurements to a distant base and most do this via the Argos satellite system (see section on *Buoy location systems* above, and Strangeways 2003). Argos also gives the location of the buoy, although for a more precise geographical location a GPS unit can be added to the sensor array. Recently the Iridium network of satellites has also started to be used as the telemetry link, in large part because it is cheaper and a link is available continuously, rather than only periodically as with Argos. The construction and specifications of drifter buoys are well described by Sybrandy and Niiler (1991) and Lumpkin and Pazos (2005).

Accuracy of the SST data from drifting buoys

Even though the majority of SST data are now coming from drifting buoys, their records go back only a short time and so cannot add to our knowledge of past SST trends, although they will be extremely important in the future. In order that their future data can be merged seamlessly with past data, it will be necessary to compare their measurements with measurements from ships. Literature analysing the errors of these buoys is, as yet, scarce compared with the detailed work concerning ship measurements; Emery et al. (2001), Reynolds et al. (2002), Rayner et al. (2006) and Smith et al. (2008) probably having done the most detailed investigations. The general wisdom, however, seems to be that the buoys provide a reasonable estimate of SST, while ships are biased warm by about 0.1–0.2°C, although this assumption has not been examined in great detail yet (J.J. Kennedy, personal communication, 2008). This would seem to be a reasonable conclusion since the buoys are permanently immersed in the sea and not subject to all the potential errors of buckets and ERIs. Their shallowness might be a problem, but as measurements are made hourly, any diurnal cycles will be detectable and indeed will be of interest and use.

References

Barlow, D. (1994) From wind stars to weather forecasts: the last voyage of Admiral Robert FitzRoy. *Weather* **49**, 123–8.

Barlow, D. (1997) The devil within: evolution of a tragedy. *Weather* **52**, 337–41.

Berry, D. I. and Kent, E. C. (2005) The effect of instrument exposure on marine air temperatures: an assessment using VOSClim data. *International Journal of Climatology* **25**(7), 1007–22.

Berry, D. I., Kent, E. C. and Taylor, P. K. (2004) An analytical model of heating errors in maritime air temperatures from ships. *Journal of Atmospheric and Oceanic Technology* **21**, 1198–215.

Chenoweth, M. (1996) Nineteenth-century marine temperature data: comments on observing practices and potential biases in marine datasets. *Weather* **51**, 280–5.

Christy, J. R., Parker, D. E., Brown, S. J., Macadam, I., Stendel, M. and Norris, W. B. (2001) Differential trends in tropical sea surface and atmospheric temperatures since 1979. *Geophysical Research Letters* **28**(1), 183–6.

Dickson, B. and Østerhus, S. (2007) One hundred years in the Norwegian Sea. *Norsk Geografisk Tidsskrift* [*Norwegian Journal of Geography*] **61**, 56–75.

Donlon, C., Robinson, I., Casey, K. S., Vazquez-Cuervo, J., Armstrong, E., Arino, O., Gentemann, C., May, D., LeBorgne, P., Piollé, J., Barton, I., Beggs, H., Poulter, D. J. S., Merchant, C. J., Bingham, A., Heinz, S., Harris, A., Wick, G., Emery, B., Minnett, P., Evans, R., Llewellyn-Jones, D., Mutlow, C., Reynolds, R. W., Kawamura, H. and Rayner, N. (2007) The Global Ocean Data Assimilation Experiment High Resolution Sea Surface Temperature Pilot Project. *Bulletin of the American Meteorological Society* **88**(8), 1197–213.

Donlon, C., Robinson, I. S., Reynolds, M., Wimmer, W., Fisher, G., Edwards, R. and Nightingale, T. J. (2008) An infrared sea surface temperature autonomous radiometer (ISAR) for deployment aboard Volunteer Observing Ships (VOS). *Journal of Atmospheric and Oceanic Technology* **25**(1), 93–113.

Emery, W. J., Baldwin, D. J., Schlüssel, P. and Reynolds, R. W. (2001) Accuracy of in situ sea surface temperatures used to calibrate infrared satellite measurements. *Journal of Geophysical Research – Oceans* **106**(C2), 2387–405.

EUMETSAT (1995) *Data Collection System, User Guide*. EUM UG 02, pp. 29–34.

Folland, C. (2005) Assessing bias corrections in historical sea surface temperature using a climate model. *International Journal of Climatology* **25**(7), 895–911.

Folland, C. K. and Parker, D. E. (1995) Correction of instrumental biases in historical sea surface temperature data. *Quarterly Journal of the Royal Meteorological Society* **121**, 319–67.

Folland, C. K., Reynolds, R. W., Gordon, M. and Parker, D. E. (1993) A study of six operational sea surface temperature analyses. *Journal of Climate* **6**(1), 96–113.
Houghton, J. (1997) *Global Warming*. Cambridge University Press.
Jones, P. D. (2008) *Temperature*. Available at www.cru.uea.ac.uk/cru/data/temperature/.
Kennedy, J. J., Brohan, P. and Tett, S. F. B. (2007) A global climatology of the diurnal variations in sea-surface temperature and implications for MSU temperature trends. *Geophysical Research Letters* **34**(5), L05712.
Kent, E. C. and Taylor, P. K. (1991) *A Catalogue of the Voluntary Observing Ships Participating in the VSOP-NA*. World Meteorological Organization Marine Meteorology and Related Oceanographic Activities Report No. 25 (WMO/TD-No. 456).
Kent, E. C. and Taylor, P. K. (1995) A comparison of sensible and latent heat flux estimates for the North Atlantic Ocean. *Journal of Physical Oceanography* **25**(6), 1530–49.
Kent, E. C. and Taylor, P. K. (2006a) *Report on the Voluntary Observing System*. Prepared for the eleventh session of the Joint GCOS-GOOS-WCRP Ocean Observations Panel for Climate, 16–20 May 2006, Ocean Research Institute, University of Tokyo.
Kent, E. C., and Taylor, P. K. (2006b) Toward estimating climatic trends in SST. Part 1. Methods of measurement. *Journal of Atmospheric and Oceanic Technology* **23**(3), 464–75.
Kent, E. C., Taylor, P. K., Truscott, B. S. and Hopkins, J. S. (1993) The accuracy of voluntary observing ships' meteorological observations: results of the VSOP-NA. *Journal of Atmospheric and Oceanic Technology* **10**, 591–608.
Kenworthy, J. M. and Walker, J. M. (eds) (1997) *Colonial Observatories and Observations: Meteorology and Geophysics*. Proceedings of a Conference of the Royal Meteorological Society held at the University of Durham, 8–10 April 1994. Occasional Publication No. 31, University of Durham, UK.
Lumpkin, R. and Pazos, M. (2005) *Measuring Surface Currents with Surface Velocity Program Drifters: The Instrument, its Data, and Some Recent Results*. NOAA, Atlantic Oceanographic and Meteorological Laboratory: Miami, FL.
Maury, M. F. (1858) *Explanations and Sailing Directions to Accompany the Wind and Current Charts: Vol. 1*. Cornelius Wendell: Washington DC.
Met Office (1995) *Marine Observer's Handbook*. Met Office 887. HMSO: London.
Met Office (1997) *UK Meteorological Office Moored Data Buoy: Technical Description*. Meteorological Office Technical Description K-27-TD, Issue 1, August 1997.
NOCS (2009) *Ocean Observing and Climate*. National Oceanography Centre: Southampton. Available at www.noc.soton.ac.uk/ooc/PROJECTS/ASMOS/.
Parker, D. E., Folland, C. K. and Jackson, M. (1995) Marine surface temperature: observed variations and data requirements. *Climatic Change* **31**(2–4), 559–600.
Rayner, N. A., Parker, D. E., Horton, E. B., Folland, C. K., Alexander, L. V., Rowell, D. P., Kent, E. C. and Kaplan, A. (2003) Global analyses of sea surface temperature, sea ice, and night marine air temperature since the late nineteenth century. *Journal of Geophysical Research* **108**(D14), 4407.
Rayner, N. A., Brohan, P., Parker, D. E., Folland, C. K., Kennedy, J. J., Vanicek, M., Ansell, T. J. and Tett, S. F. B. (2006) Improved analyses of changes and uncertainties in sea surface temperature measured in situ since the mid-nineteenth century: the HadSST2 dataset. *Journal of Climate* **19**(3), 446–69.

Reynolds, R. W., Rayner, N. A., Smith, T. M., Stokes, D. C. and Wang, W. (2002) An improved in situ and satellite SST analysis for climate. *Journal of Climate* **15**(13), 1609–25.

Roll, H. U. (1951) Water temperature measurement on deck and in the engine room. *Annals of Meteorology* **4**, 439–43.

Smith, T. M., Reynolds, R. W., Peterson, T. C. and Lawrimore, J. (2008) Improvements to NOAA's historical merged land–ocean temperature analysis (1880–2006). *Journal of Climate* **21**(10), 2283–96.

Strangeways, I. C. (2003) *Measuring the Natural Environment* (2nd edn). Cambridge University Press.

Sybrandy, A. L. and Niiler, P. P. (1991) *The WOCE/TOGA Lagrangian Drifter Construction Manual*. WOCE Report No 63; SIO Report No 91/6. Scripps Institution of Oceanography: La Jolla, CA.

Thompson, D. W. J., Kennedy, J. J., Wallace, J. M. and Jones, P. D. (2008) A large discontinuity in the mid-twentieth century in observed global-mean surface temperature. *Nature* **453** (28 May), 646–50.

Woodruff, S. D., Diaz, H. F., Elms, J. D. and Worley, S. J. (1998) COADS Release 2 data and metadata: enhancements for improvements of marine surface flux fields. *Physics and Chemistry of the Earth* **23**, 517–27.

Worley, S. J., Woodruff, S. D., Reynolds, R. W., Lubker, S. J. and Lott, N. (2005) ICOADS Release 2.1 data and products. *International Journal of Climatology* **25**, 823–42.

Yu, L. and Weller, R. A. (2007) Objectively analyzed air–sea heat fluxes for the global ice-free oceans (1981–2005). *Bulletin of the American Meteorological Society* **88**(4), 527–39.

Zhang, Y., Rossow, W. B., Stackhouse, P., Romanou, A. and Wielicki, B. A. (2007) Decadal variations of global energy and ocean heat budget and meridional energy transports inferred from recent data sets. *Journal of Geophysical Research* **112**(D22), D22101.

6

Measuring sea temperature profiles

In the previous chapter we saw how, with the development of Lagrangian drifting buoys, surface currents are now being investigated much more efficiently and with a wider geographical cover than was possible with ships. So, too, is the temperature of the sea. But these measurements are made at the surface only, perhaps to a depth of a few tens of metres at most. Currents and temperature, however, vary with depth. For a more complete understanding of the oceans, these sub-surface conditions are as important as those at the surface.

The bathythermograph

Historically, an instrument known as a *bathythermograph* was used to measure the change of temperature with depth into the sea, this taking the form of a bimetallic strip (Strangeways 2003) that recorded a trace on a smoked-glass slide, the whole being contained in a bronze, torpedo-shaped housing which was lowered into the sea by winch. After recovery, the slide was read against a graduated scale.

The expendable bathythermograph

In the 1960s, the electronic *expendable bathythermograph* (XBT) was developed to replace the mechanical bathythermograph, the temperature now being measured with a thermistor housed in an expendable casing which is allowed to sink under its own weight, the depth being estimated through knowledge of the rate at which it sinks. Readings are transferred from instrument to ship via a fine cable that spools out as the casing falls, with the cable eventually running out and breaking, the XBT sinking to the ocean floor. Airborne versions (AXBT) use radio to transmit the data to the aircraft. Some XBTs also measure conductivity, to allow salinity to be calculated.

Over 5 million XBTs have been made since the 1960s. During the 1990s around 40,000 profiles per year were collected to depths of 450 and 750 m under the TOGA (Tropical Ocean Global Atmosphere) experiment and the WOCE (World Ocean Circulation Experiment) projects, with up to an additional 20,000 profiles from other sources (particularly from TAO, the Tropical Atmosphere Ocean array).

As with buckets, ERIs and drifting buoys on the surface, sub-surface measurements from XBTs are now suspected of having biases and errors, and currently there are several research teams investigating this (Gouretski and Koltermann 2007; Willis et al. 2007).

However, these observations are limited to the shipping lanes, with few measurements south of 20°S, and this gave rise to the development of the next generation of profilers – the Argo float.

Argo: a drifting profiler float

Origins of Argo

The idea of a neutrally buoyant float for measuring sub-surface ocean currents was first put forward in the 1950s, both in the USA (Stommel 1955) and in the UK (Swallow 1955) independently. Their designs remained unchanged until the early 1970s. The floats were tracked by monitoring the continuous 10 kHz sound they emitted.

The TOGA and WOCE experiments, started in the 1980s, have shown how important the oceans are in the coupled climate system, in particular the influence that the El Niño Southern Oscillation (ENSO) and the several other 'oscillations' (North Atlantic, Arctic, Pacific Decadal, Indian Ocean Dipole and Antarctic Circumpolar Wave) have on the climate. The experiments also showed that ocean currents carry vast amounts of heat from the tropics to the mid-latitudes (Bryden et al. 1991), with large inter-annual variations of up to 30% (Roemmich and Owens 2000), the reasons for which are unknown. It is clear that global measurements of heat storage and transport in the oceans are essential in any climate observing system.

In addition to this compelling scientific reason for better measurements, innovative instrumental developments were emerging that made an improved network practical. Profiling floats were developed during the 1990s and precision satellite altimeters measuring sea level were about to be put into orbit – the two being essential partners. The Argo project emerged from two separate proposals: the first for an 'Array for Real-time Geostrophic Oceanography', named *Argo* (partly as an acronym, but also after the ship of the Greek hero Jason – emphasising the close relationship of the new float network with the new satellite-based *Jason-1* altimeter mission). The second proposal was for a programme titled the Global Ocean Salinity Monitoring (GOSAMOR) project.

In 1998 the International Steering Team for GODAE (Global Ocean Data Assimilation Experiment) undertook to develop the idea of an integrated system combining these two sources of data along with the satellite altitude measurements (Stammer and Chassignet 2000). The Upper Ocean Panel of CLIVAR (Climate Variability and Predictability) of the World Climate Research Programme (WCRP) agreed that the proposals should be given high priority. Out of this jungle of acronyms (and there are many more), and pursued by diverse influential working groups, came the Argo project.

The first ten Argo floats came into use in 1999 and the number has increased to its present total of over 3,000, now producing nearly 9,000 profiles a month. The floats are considered expendable because the cost of recovery is much greater than the cost of the float. Occasionally, however, one is recovered by chance and this provides useful information on sensor stability and ruggedness.

The Argo hardware

The floats are normally 'parked' at a depth of about 1,000 m, in sleep mode, moving with the currents at that depth, which are believed to be slow (Figure 6.1). Because of the slow movement, large gaps between – and clusters of – floats tend not to form. After ten days, they 'awake' and descend to 2,000 m (where, in many ocean regions, there is a stable temperature and salinity relationship, allowing a check on the salinity sensor's calibration to be made) and then ascend, measuring temperature and salinity as they rise, taking around six hours to reach the surface. On arrival at the surface, where they remain for 6–12 hours, their measurements are normally transmitted via the Argos system carried on polar orbiting satellites. This lengthy period at the surface is needed in order to catch a satellite overpass; the timing of overpasses varying, depending on latitude, and being least frequent near the equator, where overpasses occur about every three or four hours. Transmissions from the DCP may not always be received successfully, hence the need to stay at the surface for 6–12 hours. The location of each float is also given by the Argos system (by virtue of the satellite's movement relative to the ground). Thereafter the float sinks back down to the 1,000 m parking depth, repeating the cycle until its power is exhausted, after about four or five years (or around 180 round trips to the surface and back). A parking depth of 1,000 m is used to give a uniform single 'horizon' at which to compile a global velocity dataset.

Floats are currently made by two companies – the APEX float in the USA and PROVOR floats in France – while the SOLO float is constructed by the Woods Hole

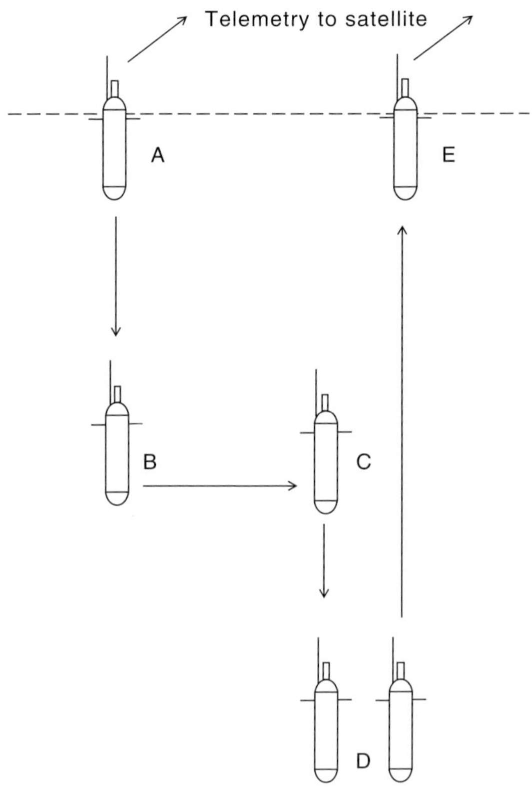

Figure 6.1 Section through the ocean showing the path taken by Argo floats. (A) Float at surface transmitting previously measured readings, (B) float descends to 1,000 m parking depth, (C) float drifts with currents for ten days, (D) float descends to 2,000 m depth and takes a reading at that depth, (E) float ascends to surface taking readings as it rises and transmits the data via satellite.

and Scripps Oceanographic Institutes. The cost of each float is around £10,000, to which must be added the cost of using the satellite. Floats are made of aluminium, weigh about 30 kg and are around 2 m long in total, including a 70 cm vertical antenna (Figure 6.2). Buoyancy is controlled by pumping oil from within the body of the float into an external bladder to send the float up, the oil being drawn back into the body of the float to sink it back down. Only about 250 g of oil is needed to move the float up and down. A recent development is the 'glider': a float with wings that allows it to glide along a programmed track as it rises or descends. Buoyancy is controllable to ±30 m depth.

Temperature is measured between –2 and +35°C to an accuracy of ±0.01°C at 2 m intervals from 0 to 500 m and at 5 m intervals from 500 to 2,000 m. This is defined as the 'target accuracy'. Accuracy has also been quoted as ±0.008°C for most sensors over a period of ten years. Looking for very small changes requires high stability and precise, absolute calibration. Compared with the drifting buoys'

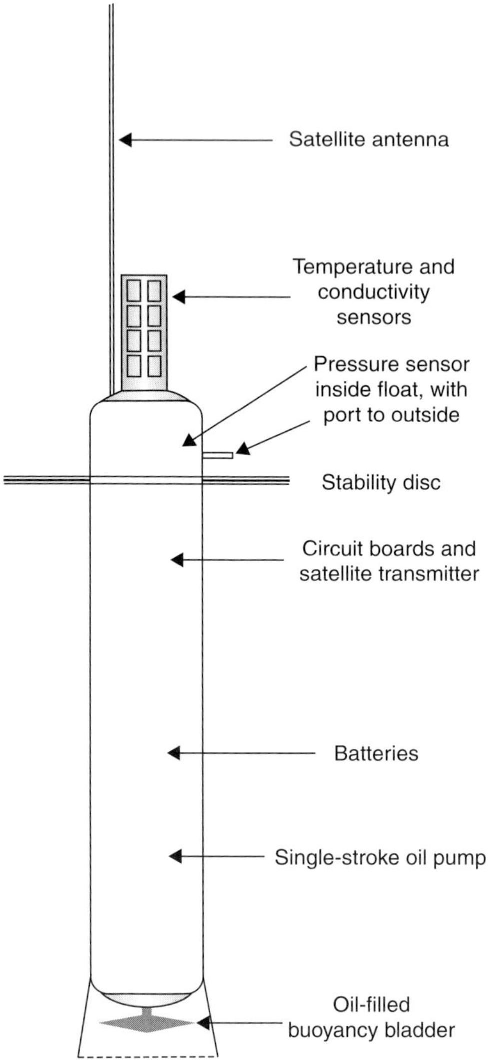

Figure 6.2 The Argo float. See text for details.

temperature accuracy of ±0.1°C, the Argo specification is an order of magnitude greater (Oka and Ando 2004).

Salinity (conductivity) is measured to 30 or 70 mS cm^{-1} to an accuracy of ±0.05 mS cm^{-1} with a resolution of 0.01 mS cm^{-1}. Pressure is measured in decibars (1 dbar = 0.1 bar), with 100 dbar being equivalent to 100 m depth of water. Argo sensors measure between 0 and 2,000 dbar (2,000 m) to an accuracy of ±5 dbar (±5 m) and a resolution of 1 dbar (1 m).

The aim was to have floats spending an hour or less at the surface to minimise biofouling of the salinity sensor, but this has not been possible because floats that use Argos communications need to be on the surface for 6–12 hours to ensure that their data are collected reliably (see above). A new Argos satellite may offer two-way communication and thus allow the float to be instructed to dive as soon as the data have been successfully received. With newer satellite systems, such as Iridium (based on a constellation of low Earth orbit (LEO) satellites), the time to transmit a full profile is typically around 15 minutes and two-way communication is also available to confirm reception and allow a return to depth. A small number of floats now use the Iridium communications system with GPS for positioning, and this is likely to increase in the coming years.

The parking depth need not be 1,000 m and can be chosen independently of profile depth, deeper parking producing less dispersion due to slower currents, as well as less biological fouling. Anti-fouling chemicals are currently used for preventing biological growth on the salinity sensor (only necessary while the float is at the surface). A further possibility is to have two sensors to help identify problems – although both are liable to suffer to the same degree! One problem in all of this is the time required to demonstrate any improvement, since tests inevitably take years.

A much shallower parking level can be selected if required; for example by placing the float at or above the thermocline (the thin dividing layer between the warm, shallow top layers of 100 m or so and the colder water beneath, across which there is a relatively large change in temperature). If placed here, the float can be dispersed more rapidly by surface currents, while also measuring conditions at the level of interest. With two-way satellite communication, the float could be instructed to go to any specified depth.

Apart from a low quiescent current in the sleep mode, power is drawn from the batteries mainly while taking readings on the six-hour journey to the surface and then making repeated transmissions to the Argos satellite every two or three minutes for 6–12 hours. Improvements have been made to the efficiency of the single-stroke pump used to inject the oil into the external bladder to initiate ascent, thereby reducing power consumption. Data are transmitted in compressed form, taking up about 2 kilobytes per profile (depending on the number of profile levels recorded).

Floats are normally launched from merchant and research ships. About half of the floats will be deployed and paid for by the USA, with other countries contributing to varying degrees. The UK, for example, via the Met Office (who manage the UK programme), the National Oceanography Centre, Southampton (NOCS), the British Oceanographic Data Centre in Liverpool and The Scottish Association for Marine Science hope to deploy up to 50 a year and maintain 150–200 operating floats at any one time (Turton and Cattle 2000). The UK has deployed around 250 floats, with about 100 currently in operation.

The UK floats have been deployed in a wide range of areas, contributing to the global array, including the North and South Atlantic, the Southern Ocean, and western and southern parts of the Indian Ocean. Turton (2002) and Gould and Turton (2006) give good overall reviews of Argo.

All Argo data are freely available via the two Global Data Assembly Centres in the USA and France, and the data are also exchanged in near-real-time via the WMO Global Telecommunications System (GTS).

Argo and Jason

Measurements of global sea level were started by NASA/CNES (CNES is the French Space Agency) using the TOPEX/Poseidon satellites in 1992 (TOPEX = Ocean Topography Experiment; in Greek mythology Poseidon was the god of the sea), measuring sea surface topography to within about ±4 cm. This was supplemented in December 2001 by *Jason-1*, a satellite-based radar altimeter that measures sea level globally every ten days to a similar accuracy as TOPEX.

If the sea was dead calm, the surface topography of the oceans would have slight peaks and troughs a few tens of centimetres in amplitude, spread over vast distances and totally invisible to the eye. These very small differences in level are due to the wind and to variations in the density of the water caused by slight differences in the water's temperature and salinity, the latter differences in level being known as the *steric* or *dynamic height*. As a column of water warms it expands. A rise of 30 cm occurred in the Eastern Pacific during the 1998 El Niño due to subsurface warming and freshening. With the arrival of Argo floats, it became possible to measure these variations in temperature and salinity in a column of water as the float rises to the surface.

Anomalies of steric height, derived from an analysis of Argo profiling buoys, have been used to assess the relative contributions of temperature and salinity to sea level changes over the North Atlantic for the period 1999–2004 (Ivchenko et al. 2007). The float measurements were compared with the dynamic topography anomalies from the Jason altimetry and it was found that the halosteric (salinity) contribution to the anomalies of dynamic height was similar in magnitude to the thermosteric (thermal) influences over the period analysed. It was found that taking both salinity and temperature into account improves the agreement between trends in the satellite dynamic topography and dynamic height, increasing the correlation between them (from 0.63 to 0.73). The implication of this (according to the authors) is that the salinity contribution cannot be neglected (in the North Atlantic at least) and that one cannot rely on estimating the thermosteric part from anomalies in the sea-surface dynamic topography derived from the satellite altimetry. In other words, it is not possible to judge sea temperatures from variations in sea topography alone; it is necessary to take salinity into account as well. Similar work was done by Gilson et al. (1998) in the North Pacific. Willis et al. (2003) describe a similar technique in the Southwest Pacific and Tasman Sea, finding a strong inter-annual variability, including the 1998 El Niño event. The authors say that the use of their method, if applied globally, would give increased insight into the variability of the general circulation and heat budget of the upper ocean.

The uses of Argo, Jason and XBT measurements

There are many specialised applications for the data collected by these instruments, ranging from brief studies of temperature and salinity beneath tropical cyclones to changes in the Arabian Sea associated with the Indian summer monsoon. On a routine basis, the Met Office runs daily forecasts of short-term deep ocean changes of temperature, salinity, currents and sea-ice in the open ocean, as well as seasonal forecasts up to six months ahead.

Our main concern, however, is with their use to determine ocean heat content, its variability and its trends. There is too short a run of data to be able to deduce much about long-term trends, but short-term variability can be studied. Even in the short term, however, there are a number of problems, in particular because of under-sampling, especially before Argo became available. The most comprehensive set of data is held by the National Oceanographic Data Center (NODC) in the USA, with sub-surface observations going back to 1947. Coverage is extremely poor, however, until the early 1970s, when XBTs began to be used routinely. Much more recently, Argo measurements have been added to the data available.

A further difficulty in estimating short-term ocean temperature variations is that low-frequency circulation changes can dominate local temperature changes. This, combined with very geographically uneven sampling, has the potential to produce large aliased signals in the time series (Palmer et al. 2007). The IPCC (2007) reports considerable spatial, inter-annual and inter-decadal variability in the natural variability of ocean heat content. Palmer et al. (2007) are working on ways of analysing the temperature signal, designed to filter out local changes.

The fluxes of heat and freshwater between the oceans and atmosphere undergo large variations due to internal variability and external changes of climate. Direct measurements of the fluxes of heat and freshwater are difficult to make and are liable to have very large errors but, by measuring the temperature and salinity of the oceans, the overall integrated effects of the fluxes can be estimated. This is similar to the situation in the atmosphere, where it is difficult to measure water vapour fluxes (evaporation) at any one point accurately, coupled with the great spatial variability of evaporation, whereas measurement of the bulk humidity of the air is more easily achieved, giving an indirect indication of the latent heat flux. The same is true of measuring the sensible heat flux compared with the greater ease of observing bulk air temperatures.

Thus, in the case of the oceans, rather than attempt the impossible task of measuring the amount of heat and rainfall (or melting ice) entering the oceans across their entire surface, Argo can measure the end result – the temperature and salinity of the bulk of the ocean, and make inferences from them. The UK Hadley Centre have recently used data from XBT and Argo for this purpose (Banks et al. 2000; Gregory et al. 2004; Wu et al. 2004) but found that the limited sampling in the past makes it difficult to attribute any of the observed changes to either natural or anthropogenic causes. Most historical ocean observations are from the Northern Hemisphere and, as Harrison and Carson (2007) point out, there are no data for most of the ocean south of 20°S. From the limited data available, they conclude that the oceans would appear neither to have cooled nor warmed systematically over the period 1950–2000 (not at the surface but over the depth of measurements made).

The greatly increased availability of data that will result from Argo, however, will allow a better understanding of the variability of the climate system and will enable model projections of future change to be refined. Argo offers the possibility of learning how the 'state of the oceans' changes in the *coming* decades and will throw light on large-scale interactions between the oceans and the atmosphere. For this to occur, it is essential that the Argo array be completed – and be maintained into the future. Observations from the real world are essential for our understanding of the climate. Instrument measurements in larger numbers have only just started to be made in the ocean depths and they are not yet sufficient to make confident statements about trends, although information about short-term variability of ocean heat uptake is just beginning to emerge (Achuta Rao et al. 2006).

References

Achuta Rao, K. M., Santer, B. D., Gleckler, P. J., Taylor, K. E., Pierce, D. W., Barnett, T. P. and Wigley, T. M. L. (2006) Variability of ocean heat uptake: reconciling observations and models. *Journal of Geophysical Research – Oceans* **111**, C05019.

Banks, H. T., Wood, R. A., Gregory, J. M., Johns, T. C. and Jones, G. S. (2000) Are observed decadal changes in intermediate water masses a signature of anthropogenic climate change? *Geophysical Research Letters* **27**(18), 2961–4.

Bryden, H. L., Roemmich, D. H. and Church, J. A. (1991) Ocean heat transport across 24°N in the Pacific. *Deep-Sea Research* **38**, 297–324.

Gilson, J., Roemmich, D., Cornuelle, B. and Fu, L.-L. (1998) Relationship of TOPEX/Poseidon altimetric height to steric height and circulation in the North Pacific. *Journal of Geophysical Research* **103**, 27947–65.

Gould, W. J. and Turton, J. (2006) Argo: sounding the oceans. *Weather* **61**, 17–21.

Gouretski, D. and Koltermann, K. P. (2007) How much is the ocean really warming? *Geophysical Research Letters* **34**, L01610.

Gregory, J. M., Banks, H. T., Stott, P. A., Lowe, J. A. and Palmer, M. D. (2004) Simulated and observed decadal variability in ocean heat content. *Geophysical Research Letters* **31**, L15312.

Harrison, D. E. and Carson, M. (2007) Is the world ocean warming? Upper-ocean temperature trends: 1950–2000. *Journal of Physical Oceanography* **37**, 174–87.

IPCC (2007) *Climate Change 2007: The Physical Science Basis*. Contribution of Working Group I to the Fourth Assessment Report of the Intergovernmental Panel on Climate Change. Cambridge University Press.

Ivchenko, V. O., Danilov, S. D., Sidorenko, D. V., Schröter, J., Wenzel, M. and Aleynik, D.L. (2007) Comparing the steric height in the northern Atlantic with satellite altimetry. *Ocean Science* **3**(4), 485–90.

Oka, E. and Ando, K. (2004) Stability of temperature and conductivity sensors of Argo profiling floats. *Journal of Oceanography* **60**(2), 253–8.

Palmer, M. D., Haines, J., Tett, S. F. B. and Ansell, T. J. (2007) Isolating the signal of ocean global warming. *Geophysical Research Letters* **34**, L23610.

Roemmich, D. and Owens, W. B. (2000) The Argo project: global ocean observations for understanding and prediction of climate variability. *Oceanography* **13**, 45–50.

Stammer, D. and Chassignet, E. P. (2000) Ocean state estimation and prediction in support of oceanographic research. *Oceanography* **13**, 51–6.

Stommel, H. (1955) Direct measurement of subsurface currents. *Deep-Sea Research* **2**(4), 284–5.

Strangeways, I. C. (2003) *Measuring the Natural Environment* (2nd edn). Cambridge University Press.

Swallow, J. C. (1955) A neutral-buoyancy float for measuring deep currents. *Deep-Sea Research* **3**(1), 93–104.

Turton, J. (2002) Argo 2002: progress towards a global array of profiling floats. Paper presented at *Oceanology International 2002*, ExCel, London.

Turton, J. and Cattle, H. (2000) The UK contribution to Argo. *NWP Gazette* March 2000, 6–7.

Willis, J. K., Roemmich, D. and Cornuelle, B. (2003) Combining altimetric height with broadscale profile data to estimate steric height, heat storage, subsurface temperature and sea-surface temperature variability. *Journal of Geophysical Research* **108**(C9), 3292.

Willis, J. K., Lyman, J. M., Johnson, G. C. and Gilson, J. (2007) Correction to 'Recent cooling of the upper ocean'. *Geophysical Research Letters* **34**, L16601.

Wu, P., Wood, R. and Stott, P. (2004) Does the recent freshening trend in the North Atlantic indicate a weakening thermohaline circulation? *Geophysical Research Letters* **31**, L02301.

7

Global instrument networks

While innumerable measurements of temperature have been made since the thermometer was invented, many are not suitable for following changes in climate, either because the records are too short in duration or because they do not meet basic technical requirements. If we want to look at past climates, going back a century or more, there is no recourse other than to draw upon data that originate from traditional, manually operated mechanical instruments as described in earlier chapters. Indeed many of today's data still originate from this type of instrument. Most of the suitable stations are to be found amongst the long-established, existing networks administered by the world's National Meteorological Services (NMSs). In this chapter we look at the networks that are being used to follow climate change – in particular temperature.

First station compilations

Attempts to investigate temperature change started as long ago as the late nineteenth century, when Köppen (1881) examined the relatively few measurements then available, from 1840 to 1880, using land air temperatures from the tropics and temperate latitudes. But it was another 50 years before the next attempt was made, this time by Callendar (1938), who collected data from land stations covering the whole globe for the period 1889–1938. Twelve years later, Willett (1950) assembled data from 1845 to 1945 from global land stations. Callendar (1961) then extended his run of data back to 1870 over the range from 60°N to 60°S from the land surfaces. Mitchell (1961, 1963) was the next, collecting data from a global network of land stations starting around 1880. There were a number of papers published in the 1970s, but those by the Russian authors Budyko (1969), updated by Borzenkova et al. (1976), and by Vinnikov et al. (1980) were the most detailed. Their data, which covered just the Northern Hemisphere, were drawn from three series of charts processed by the Main Geophysical Observatory (GGO), published as monthly average temperatures for 1881–1935. Other charts, for 1961–75, were published in the Synoptic Bulletin of the USSR Hydrometeorological Centre. The temperature trends brought to light by these early workers are illustrated in the IPCC's Fourth Assessment Report (AR4, IPCC 2007, p. 101).

Because each study covered a different time period and did not use identical data, they are not directly intercomparable, but despite this they show surprisingly good agreement, even with today's estimates. It was found by Jones (1994) that even a limited number of temperature records can give reliable global trends over a year. This finding will be returned to in the final chapter when the design of a new instruments network for the future is discussed.

World Weather Records (WWR)

In the 1970s the main sources of long-term global meteorological records were contained in the World Weather Records (WWR) published by the Smithsonian Institution (Smithsonian Institution 1927, 1935, 1947) and the US Weather Bureau (US Weather Bureau 1959–1982; Vose et al. 1992). In addition there are similar publications for each subsequent decade up to the present. These records were digitised by the National Center for Atmospheric Research (NCAR) (Jenne 1975). Monthly values from 1961 onwards have been published by the National Oceanic and Atmospheric Administration (NOAA) in their Monthly Climatic Data for the World (MCDW).

The Climatic Research Unit (CRU) network

A significant step towards improving on the work done in Russia and on the WWR was reported by Jones et al. (1982), who added many new stations, data being derived from the published archives of the National Meteorological Services, filling in gaps in the WWR coverage. Data were also supplied by the Danish and Icelandic Met Services, with additional data from Norway, the USSR and from the UK Met Office. Better coverage could have been obtained, but only through contacting each NMS individually and directly, which would have been very slow and not necessarily successful.

A few years later, funded by the US Department of Energy, Carbon Dioxide Research Division, Jones et al. (1985) produced an updated compilation of stations for the Northern Hemisphere comprising monthly surface air temperatures, the aim being to produce a gridded dataset for the hemisphere. (How the homogeneity of the records was checked and how the gridded data were produced from individual station records is the topic of the next chapter.) Considerable additional data were digitised from the yearbooks of the National Meteorological Services and from many other sources. Of the 2,666 stations considered, 1,584 were deemed to produce good enough data to be included in a new dataset (Bradley et al. 1985). A year later, a similar compilation was made for the Southern Hemisphere, comprising 293 stations drawn from a possible 610 (Jones et al. 1986).

This apparent dearth of data from the Southern Hemisphere partly reflects the fact that there is much less land in the south, especially if Antarctica is excluded. If the USA and Europe are excluded in the north, the numbers are comparable for the Northern and Southern Hemispheres.

The Global Historical Climatology Network (GHCN)

Until recently the largest directory of stations had been that assembled by the National Climatic Data Center (NCDC), operated by the National Oceanic and Atmospheric Administration (NOAA) and the Carbon Dioxide Information Analysis Center (CDIAC) at Oak Ridge National Laboratory in Tennessee. Known as the *Global Historical Climatology Network* (GHCN), the data are collected from about 6,000 temperature stations (along with 7,500 precipitation and 2,000 barometric pressure stations). Data from these stations are used to construct past global temperatures, such as NASA's 'GISTEMP' (Goddard Institute for Space Studies, GISS). The average temperature record is 60 years long with about 1,650 records longer than 100 years and 220 longer than 150 years.

The Global Climate Observing System (GCOS)

The most recent of the instrument networks is the GCOS, set up by the WMO, co-sponsored by the Intergovernmental Oceanographic Commission (IOC) of UNESCO, the United Nations Environment Programme (UNEP) and the International Council for Science Unions (ICSU).

In the WMO's own words (see:
www.wmo.int/pages/prog/gcos/documents/GCOS_brochure_GoodQuality.pdf,
www.wmo.int/pages/prog/gcos/index.php?name=AboutGCOS):

> The Global Climate Observing System (GCOS), established in 1992, is an internationally coordinated system of observing systems and networks for meeting national and international needs for climate observations. It consists of the climate-relevant components of all established environmental observing networks and systems, and it serves as the climate observation component of the Global Earth Observation System of Systems (GEOSS).
>
> GCOS stimulates, encourages, coordinates and otherwise facilitates the taking of the needed observations by national or international organizations in support of their own requirements as well as of common goals. It provides an operational framework for integrating, and enhancing as needed, observational systems of participating countries and organizations into a comprehensive system focussed on the requirements for climate issues. The

GCOS programme does not directly make observations nor generate data products. GCOS builds upon, and works in partnership with, other existing and developing observing systems such as the WMO Global Observing System and Global Atmosphere Watch, the Global Ocean Observing System, and the Global Terrestrial Observing System. It includes in situ, airborne and space-based observational components. GCOS builds, to the extent possible, on existing observing, data management and information distribution systems, both operational and research, and on further enhancements of these systems. These include:

- **WMO Global Observing System** for atmospheric physical and dynamical properties
- **WMO Global Atmosphere Watch** for atmospheric constituent and chemical properties
- **ICSU/IOC/WMO Global Ocean Observing System** for physical, chemical and biological properties
- **FAO/ICSU/UNEP/WMO Global Terrestrial Observing System** for land surface ecosystem, hydrosphere and cryosphere measurements
- **WCRP GEWEX Baseline Surface Radiation Network** for surface radiation budget measurements
- **IGBP** and **WCRP** research networks, for monitoring of terrestrial ecosystems, clouds and the hydrological cycle, the Earth's radiation budget, ice sheets and precipitation over the oceans
- Data communication and other infrastructures necessary to support operational climate forecasting, including the **World Climate Data and Monitoring Programme** (WCDMP) and the **Climate Information and Prediction Services** (CLIPS).

From this it will be seen that the GCOS encompasses virtually all aspects of environmental measurements. Our interest here focuses on land and sea temperature stations.

Land surface air temperature stations

The part of the GCOS that deals with the land surface network is made up of stations drawn from four sources (Peterson et al. 1997): the Global Historical Climatology Network (see above), the Climatic Research Unit (CRU) dataset (see above), WMO 1961–90 '*Normals Stations*' (stations that produce averages for that 30-year period; 3,342 stations) and the list of *Reference Climatological Stations* (RCSs) – specially selected (by the NMS), high-quality stations with a long history of observations (2,283 stations).

According to WMO recommendations, each Member Country should operate at least one RCS, but this request tends to be interpreted widely, with some countries including all of their major sites in all of their networks (sometimes for political/economic reasons), some only a few, while others do not respond at all. Stations from these four sources, totalling 15,433, were sifted to avoid duplication, and then ranked according to:

- Length of record
- Data quality (homogeneity – defined in Chapter 8)

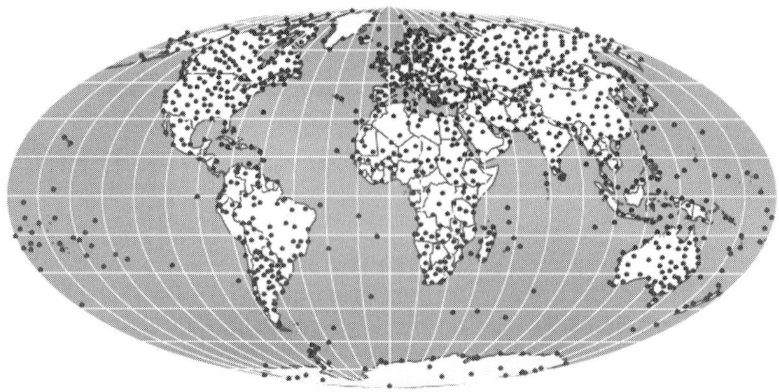

Figure 7.1 Map showing the location of the 1,023 GCOS stations in 2008 (reproduced with the kind permission of the WMO).

- Whether the station was an RCS
- Whether the station reported in real-time over the GTS (see section 'Disseminating the data' below)
- The size of the surrounding population – to minimise urban heat island (UHI) effects
- Good prospects of the site continuing to operate into the future
- Probability of the site remaining substantially unchanged over time.

Having rated the stations according to these criteria, they were further selected depending on their geographical position and how this affected the calculation of gridded means (see Chapter 8). The final selection is made up of just over 1,000 stations (Figure 7.1) (see www.gosic.org/gcos/GCOS-dev.htm (click on 'List of GSN Stations') and www.wmo.int/pages/prog/gcos/documents/GSN_Station_Map.pdf).

GCOS ocean networks

The GCOS encompasses all of the existing ocean instrument networks, as discussed in the previous two chapters, with most ocean data coming from ships and, more recently, from fixed and drifting buoys, while Argo floats now produce profiles throughout the depth of the oceans.

The Ten Monitoring Principles of the GCOS

The aim of the GCOS was first and foremost to identify and select existing instrumental networks from which data could be used for the purposes of monitoring the climate and its changes. With regard to land surface temperatures, the GCOS was, in effect, following in the footsteps of the work already pioneered and developed by the other organisations and individual workers through the period 1950–90, as described above, indeed it drew heavily on these. In this, the GCOS has been very successful. The task of the GCOS has been one of cataloguing existing stations and creating a directory of instrument sites, with metadata where available.

Being aware of the shortcomings of the old and existing instrument networks comprising the land surface section of the GCOS, the WMO has been proactive in drawing up a list of *Ten Monitoring Principles* of the GCOS, compiled to guide their improvement. These principles are, in the GCOS's own words:

Effective monitoring systems for climate should adhere to the following principles:

1. The impact of new systems or changes to existing systems should be assessed prior to implementation.
2. A suitable period of overlap for new and old observing systems should be required.
3. The results of calibration, validation and data homogeneity assessments, and assessments of algorithm changes, should be treated with the same care as data.
4. A capacity to routinely assess the quality and homogeneity of data on extreme events, including high-resolution data and related descriptive information, should be ensured.
5. Consideration of environmental climate-monitoring products and assessments, such as IPCC assessments, should be integrated into national, regional and global observing priorities.
6. Uninterrupted station operations and observing systems should be maintained.
7. A high priority should be given to additional observations in data-poor regions and regions sensitive to change.
8. Long-term requirements should be specified to network designers, operators and instrument engineers at the outset of new system design and implementation.
9. The carefully planned conversion of research observing systems to long-term operations should be promoted.
10. Data management systems that facilitate access, use and interpretation should be included as essential elements of climate monitoring systems.

The above are the main principles, to which the WMO/GCOS has appended the following points concerning satellite observations:

Furthermore, satellite systems for monitoring climate need to:

(a) take steps to make radiance calibration, calibration-monitoring and satellite-to-satellite cross calibration of the full operational constellation a part of the operational satellite system; and
(b) take steps to sample the Earth system in such a way that climate-relevant (diurnal, seasonal and long-term interannual) changes can be resolved.

Thus satellite systems for climate monitoring should adhere to the following specific principles:

11. Constant sampling within the diurnal cycle (minimizing the effects of orbital decay and orbit drift) should be maintained.

12. A suitable period of overlap for new and old satellite systems should be ensured for a period adequate to determine inter-satellite biases and maintain the homogeneity and consistency of time-series observations.
13. Continuity of satellite measurements (i.e., elimination of gaps in the long-term record) through appropriate launch and orbital strategies should be ensured.

The ten basic principles were adopted by the Conference of the Parties (COP) to the United Nations Framework Convention on Climate Change (UNFCCC) through decision 5/CP.5 at COP-5 in November 1999. The complete set of principles was adopted by the Congress of the World Meteorological Organization (WMO) through Resolution 9 (Cg-XIV) in May 2003; agreed by the Committee on Earth Observation Satellites (CEOS) at its 17th Plenary in November 2003; and adopted by COP through decision 11/CP.9 at COP-9 in December 2003 (see www.wmo.ch/pages/prog/gcos/index.php?name=ClimateMonitoringPrinciples and www.wmo.int/pages/prog/gcos/Publications/gcos-92_GIP_ES.pdf).

This is well thought-through, well planned and thoroughly documented. The list of actions is entirely correct. If these aims could be achieved we would be well set to monitor climate change much better in the future than in the past or the present. There is, however, a 'But'.

Because it is up to individual countries whether they have a mind to, can afford to, or are able to modify existing stations and procedures to meet the new standards, the principles are slow in being adopted in many parts of the world. The aims are indeed very ambitious, perhaps unrealistically idealistic, and so probably impracticable for many sites. Concerns have been expressed by the WMO about whether success will be achieved: in the *Second Adequacy Report* (see www.wmo.ch/pages/prog/gcos/Publications/gcos-82_2AR.pdf) it was stated:

The Report concludes, in agreement with the IPCC, that there remain serious deficiencies in the ability of the current global observing systems for climate to meet the observational needs of the UNFCCC. The Report in its various findings documents the needs and opportunities for improvement to the global observing systems for climate. Without urgent action to address these findings, the Parties will lack the information necessary to effectively plan for and manage their response to climate change. It requires immediate action by the Parties, the UNFCCC and intergovernmental and international agencies, and will require the allocation of additional resources.

In conversation with some of those involved with the GCOS, there would appear to be frustration at the slow rate of progress being made towards improving the existing network of stations to meet the Ten Monitoring Principles of the GCOS. In many developing countries, however, there is not the money, the expertise, the skills, the specialised knowledge or the ability to act in a timely fashion. I have worked in many countries on instrument projects and have experienced these problems at first hand. Aid helps a little, but the problems cannot be solved by money alone; it goes deeper than simple cost – it is often economic, cultural and political. I will suggest in the final chapter what our best options are for the future with regard to measuring surface air temperatures. In the meantime one can only wish the GCOS well. The Ten Monitoring Principles are absolutely right and the intent admirable. The outcome is somewhat less certain.

I am particularly concerned at the moment about how AWSs are being introduced without international standardisation. We will end up with many different AWS designs that are difficult to intercompare, and some may not even attain the necessary standard. We will be back to making 'corrections' to the measurements. This must not be allowed to happen.

The adequacy of the GCOS network

There is another difficulty with the GCOS instrument network, quite apart from the problems discussed above regarding the upgrading of existing stations to meet the more stringent requirements of climate monitoring. This problem is not, however, unique to the GCOS network; it applies to all old manual, land-based stations – which make up nearly all of the networks. This problem concerns the geographical cover of the stations around the globe and also where the stations are positioned within countries.

Global coverage

Where stations are operated around the world today is dictated by the past and by how each individual NMS acts. In consequence, station cover is patchy, with high concentrations in Europe and the USA, and fewer in other parts of the world; the number in any one country also varies with time. Indeed the network of stations is in a constant state of flux, with stations operating perhaps for several decades before being discontinued or moved to new locations (often without any metadata to record the changes). These continual changes create problems for researchers attempting to generate datasets covering long periods of time.

But there are also many countries that have more data than they make available. While the USA and Canada have digitised almost all of their data and make it freely available, most other countries do not. There is a huge amount of data 'out there' if only it could be accessed. Of course this does not mean it would all be good data. And, for obvious reasons, coverage is poor in all regions where there are no people.

Data quality

Even where stations do exist, there can be no guarantee how good their data are and no simple way of finding out. In the next chapter, when we look at how data from individual stations around the globe are combined to give an overall picture of temperature trends, the question of checking data quality, of station distribution and the effect of the ever-changing number of stations are considered.

Local microclimates

Since most usable stations are manual, they need operators and so are invariably sited in or near towns. From the mid twentieth century, many stations have also been sited at, or moved to, airports, taking advantage of the presence of reliable operators day and night, every day of the year. Airports were also appropriate since meteorological data collected there are useful for local aviation management, as well as for general weather forecasting. Except for the few large international airports, most are relatively small and, of necessity, in open flat countryside – which is good for instrument exposure. While an airport is often suspected of affecting temperature readings due to the presence of runways, buildings and roads, this may not often be the case. Provided that the airport is small and the temperature screens are exposed over natural ground (grass, rock, sand, snow or whatever is natural at the site) and are not sited too close to buildings and aircraft, the effects of the airport infrastructure may well be negligible.

Nevertheless, it is necessary to ask whether the airports or towns have grown since the stations were installed at them. It will be recalled from Chapters 3 and 4 that it is *changes* of UHI biases that are relevant in climatology. If the site changes substantially, then the temperature readings are also likely to change as a result, and this will introduce an artificial trend into the readings which is almost impossible to detect and correct for. A tenth of a degree increase might be thought of as negligible, but it is very significant in climatological terms. How such changes can be prevented in the future is discussed in the final chapter.

Site representativeness

An equally important problem with using airports as sites for detecting climate change is that their locations are often not typical of the country which they are intended to represent. I will use Papua New Guinea (PNG) as an example, having worked there on WMO projects concerning instrument networks for water resources. There are three GCOS stations representing PNG, the first at the international airport in the capital, Port Moresby, 6 km from the south coast, at an altitude of 48 m (Figure 4.3c). The site is probably sufficiently far from the runways to be unaffected by them, but the city is encroaching upon it, making the area increasingly urban in nature. The climate is also quite different from that a few tens of kilometres inland to the north, which is largely tropical rainforest.

The second GCOS site is at Madang, a port town with a population of 20,000 on the northern coast of PNG, the site being just 5 m above sea level. Depending on wind direction, this will give measurements either of the lowland coastal air temperature or the marine air temperature.

The third GCOS station is at the small airport of Momote, on a promontory at the very eastern tip of the small island of Manus, 350 km north of PNG in the Bismarck Sea. The station is only 4 m above sea level and will be measuring marine air temperature at all times.

While these sites were perfectly acceptable for their original purposes, and continue to be so for local aviation and weather forecasting today, they are not representative of the island of New Guinea. Much of the island is covered in tropical rainforest, with mountains up to 4,000 m altitude with high rainfall and even snowfall on the mountain peaks. Many of the instrument sites I visited in the interior would be much more representative. The question arises, therefore, as to whether the temperature trends measured at these lowland sites, close to the sea, represent well enough the temperature trends across the whole of PNG. Would different trends have been observed if the stations had been at forested, inland sites?

The trends at the three stations will, of course, tell us what the trends were at the three stations and that is better than nothing. It might be argued that this is sufficient. Perhaps stations can be sited anywhere. Why do we need to know what the trends were inland? They might be the same anyway. Do places have different climate trends because they have different climates? We look more closely at these, and similar fundamental questions, in the next two chapters.

While PNG is just one example, the same situation is common globally, with many GCOS stations at airports. In the UK, for example, three of the six GCOS stations are at airports and four of the six are by the sea (Strangeways 2009).

Using *Google Earth* to assess GCOS site conditions

Google Earth

Because of the problems of microclimate and representativeness just discussed, it would help if the sites could be examined. Researchers have to use measurements from sites they have never seen and which generally have no, or little, metadata. How this has been coped with so far is discussed in the next chapter, but with the advent of *Google Earth* it became possible to look at any point on Earth in considerable detail, anyone with a PC being able to download the software and access the images free of charge (see http://earth.google.com/intl/en/userguide/v4/). The system uses images collected from satellites and aircraft joined together seamlessly. It has endless uses, from TV news reports to looking at your prospective holiday villa to ensure it is not next to a rubbish tip. Here we consider its use for looking at

GCOS (and similar) station sites in order to evaluate their local microclimate and to judge their general representativeness of the surrounding country.

Evaluating instrument sites from Google Earth *images*

While a snapshot picture of a station as it is now (or was recently) will not tell us anything about its long-term history, there will be clues regarding its present state of upkeep as well as inferences about its past, such as the presence of trees or buildings around the site, which may have grown and increased with time. The extent to which a site is embedded in a city or airport, and its exact whereabouts within them, are all valuable pieces of additional information to anyone using the data from the sites for climate research. The recent availability of *Google Earth* allows such information to be gathered very easily. (Note that *Google Earth* images are not in real-time and may be as much as several years old.)

Unfortunately, the positions of the GCOS sites are only specified in the WMO literature to one minute-of-arc resolution, this low resolution locating stations to within no better than about one kilometre of their actual position. While this is adequate for assessing how well the site represents the surrounding country, the coordinates need to be specified to the nearest second-of-arc (or to the fourth decimal place if given in decimal degrees) in order to locate the actual met enclosure itself. The WMO is thus urged to publish locations of the GCOS sites to this higher resolution – not a difficult task, given the wide availability of cheap GPS navigational aids. However, as with all such matters (for example the Ten Monitoring Principles of the GCOS), response from different countries will be very patchy and we should not expect to have the locations to the necessary resolution from most countries any time soon.

Because of complex copyright regulations (sometimes three organisations having to be approached for permission to use a single image), no *Google Earth* images are reproduced here, but examples are given (in colour) in an article that looks into the use of the images, using the six GCOS sites in the UK as examples (Strangeways 2009). The positions of UK sites are located more accurately by the Met Office through their use of National Grid coordinates, allowing the actual met enclosures to be seen.

Disseminating the data

Finally a word about how data from the GCOS sites are made available to users. Until the last few decades, all meteorological data were recorded in handwritten form (many still start out in this way, in the field), some being recorded as a pen trace on a paper chart (which then have to be digitised). Archiving was by means of handwritten, typed or printed tables and paper charts. Dissemination was by post. In the 1920s, radio telegraphy enabled data to be disseminated rapidly by Morse code, especially useful from ships. In the 1960s, data loggers were developed as digital microelectronics became available, initially using magnetic tape as the recording medium, and later electronic (RAM) memory. With data now in electronic form it became possible to transmit them by telephone, land-based radio and later by satellite (Strangeways 2003).

With this technology available, it was possible for the *Global Telecommunications System* (GTS) to be organised (by the WMO), this being a complex network connecting all of the NMSs together for the interchange of data in near-real-time for weather forecasting. The data are transmitted in WMO codes, of which there are several, including SYNOP (surface synoptic observations): a code used for reporting weather observations made by manned and automated weather stations. Reports are typically sent every six hours. BUFR is a more modern code, of which there are many versions. To receive any of these data one has to be an 'authorised user', which might prevent some individuals from accessing them. For those wanting more information on accessibility, and on the GTS generally, there are many WMO websites. As regards future plans see the WMO (www.wmo.int/pages/prog/drr/events/kobe/documents/Ichijo.pdf), which gives details of the Future WMO Information Service (FWIS).

However, not all data are distributed via the GTS, with many still appearing only in printed format, published by the WMO once a decade. Such prolonged delays cause bottlenecks in producing complete climate datasets. All climate data need to be available in near-real-time.

In the next chapter we move on to look at how the temperature measurements collected by the GCOS, and by the other instrument networks, are converted from individual point measurements into means, or averages, covering larger areas, up to the global.

References

Borzenkova, I. I., Vinnikov, K. Y., Spirana, L. P. and Stekhnovskii, D. I. (1976) Changes in the air temperature of the Northern Hemisphere for the period 1881–1975 [in Russian]. *Meteorologiya i Gidrologiya* **7**, 27–35.

Bradley, R. S., Kelly, P. M., Jones, P. D., Diaz, H. F. and Goodess, C. M. (1985) *A Climate Data Bank for the Northern Hemisphere's Land Areas, 1851–1980.* DoE Technical Report No. TR017, US Department of Energy, Carbon Dioxide Research Division, Washington DC.

Budyko, M. I. (1969) The effects of solar radiation variations on the climate of the Earth. *Tellus* **21**, 611–19.

Callendar, G. S. (1938) The artificial production of carbon dioxide and its influence on temperature. *Quarterly Journal of the Royal Meteorological Society* **64**, 223–37.

Callendar, G. S. (1961) Temperature fluctuations and trends over the Earth. *Quarterly Journal of the Royal Meteorological Society* **87**(371), 1–12.

IPCC (2007) *Climate Change 2007: The Physical Science Basis.* Contribution of Working Group I to the Fourth Assessment Report of the Intergovernmental Panel on Climate Change. Cambridge University Press.

Jenne, R. L. (1975) *Data Sets for Meteorological Research.* NCAR Technical Note NCAR-TN/JA-111. National Center for Atmospheric Research: Boulder, CO.

Jones, P. D. (1994) Hemispheric surface air temperature variations: a reanalysis and an update to 1993. *Journal of Climate* **7**, 1794–802.

Jones, P. D., Wigley, T. M. L. and Kelly, P. M. (1982) Variations in surface air temperature. Part 1. Northern Hemisphere, 1881–1990. *Monthly Weather Review* **10**(2), 59–70.

Jones, P. D., Raper, S. C. B., Santer, B., Cherry, B. S. G., Goodess, C. M., Kelly, P. M., Wigley, T. M. L., Bradley, R. S. and Diaz, H. F. (1985) *A Grid Point Surface Air Temperature Data Set for the Northern Hemisphere.* US Department of Energy, Carbon Dioxide Research Division: Washington DC.

Jones, P. D., Raper, S. C. B., Cherry, B. S. G., Goodess, C. M. and Wigley, T. M. L. (1986) *A Grid Point Surface Air Temperature Data Set for the Southern Hemisphere.* US Department of Energy, Carbon Dioxide Research Division: Washington DC.

Köppen, W. (1881) Über Mehrjährige Perioden der Wittering. III. Mehrjährige Andrrungen der Tempertur 1841 bis 1875 in den Tropen der nördlichen und südlichen gemässigten zone, an den jahresmitteln untersucht. *Zeitschrift der Österreichischen Gesellschaft für Meteorologie* **XVI**, 141–50.

Mitchell, J. M. (1961) Recent secular changes of global temperature. *Annals of the New York Academy of Sciences* **95**, 235–50.

Mitchell, J. M. (1963) On the world-wide pattern of secular temperature change. In: *Proceedings of the UNESCO/WMO Symposium on Changes of Climate.* UNESCO Arid Zone Research Series 20. UNESCO: Paris, pp. 161–81.

Peterson, T., Daan, H. and Jones, P. (1997) Initial selection of a GCOS surface network. *Bulletin of the American Meteorological Society* **78**, 2145–52.

Smithsonian Institution (1927, 1935, 1947) *World Weather Records.* Smithsonian Institution Miscellaneous Collections 79, 90, 104. Smithsonian Institution: Washington DC.

Strangeways, I. C. (2003) *Measuring the Natural Environment* (2nd edn). Cambridge University Press.

Strangeways, I. C. (2009) Using *Google Earth* to evaluate GCOS weather station sites. *Weather* **64**, 4–8.

US Weather Bureau (1959–1982) *World Weather Records 1941–50, 1951–60, 1961–70* (Volumes 1–6). US Department of Commerce: Washington DC.

Vinnikov, K. Ya., Gruza, G. V., Zakharov, V. F., Kirillov, A. A., Kovyneva, N. P. and Ran'kova, E. Ya. (1980) Contemporary variations of the northern hemisphere climate [in Russian]. *Meteorologiya i Gidrologiya* **6**, 5–17.

Vose, R. S., Schmoyer, R. L., Steurer, P. M., Peterson, T. C., Heim, R., Karl, T. R. and Eischeid, J. K. (1992) *The Global Historical Climatology Network: Long-Term Monthly Temperature, Precipitation, Sea Level Pressure, and Station Pressure Data.* ORNL/CDIAC-53, NDP-041. Available at http://cdiac.esd.ornl.gov/epubs/ndp/ndp041/ndp041.html.

Willett, H. C. (1950) Temperature trends of the past century. In: *Centenary Proceedings of the Royal Meteorological Society.* Royal Society: London, pp. 195–206.

8

From point measurements to global averages

So far we have looked at how temperature measurements are made on land with thermometers in screens, and how sea surface and marine air temperatures are measured from ships and buoys. The next step is to see how these measurements are converted into daily and then into monthly averages at individual stations, and then how average values representing an area are calculated. (The words 'average' and 'mean' are interchangeable.) Finally the ways in which hemispheric and global averages can be estimated are examined. It makes things clearer to divide this investigation into three sections: means from over the land, from over the oceans, and how land measurements can be combined with those from the sea to give a global figure.

We are moving away now from instruments and into questions regarding the use of the data collected by the instruments and this takes us into the statistical treatment of the measurements. I do not adopt a mathematical approach in dealing with these questions as this would be unintelligible to all except the specialist reader, and I would like the book to be read much more widely than a detailed mathematical treatment would permit or encourage. Instead, I express the statistical concepts in descriptive terms, while at the same time including references to papers that deal with the mathematics for those who want, and are able, to follow the details. Very occasionally some simple maths is essential, as in the next section.

Data from the land

Most of the methods used to process land data also apply to data from the oceans, but there are differences and so it is best to look at the two separately.

Working in averages

Calculating the average

Surface air temperature over the land can be defined as the temperature measured in a Stevenson-type screen (or the equivalent miniature screen as used on AWSs) at a height of around 1.5 m. The measurements used to calculate land air temperature trends and variability over time are based on daily mean temperatures from each individual station. As explained in Chapter 2, the daily mean air temperature at a land site is most usually obtained by taking the average of the daily maximum and minimum temperatures. This obviously does not give a very precise mean for each individual day because the maximum might be quite short-lived under some conditions and long-lived under others. However, over time, even over just a month, it probably gives a fair estimate. But there are also many other ways of measuring the mean, based on several spot readings made through the day and night. The following gives an idea of the many methods used. Those allergic to even simple arithmetic can skip this part.

(Max + Min)/2 (the mean of the maximum and minimum)
(08 + 15)/2 (where 08 and 15 are readings taken at those hours)
(07 + 13 + 19)/3
(07 + 14 + 21)/3
(00 + 06 + 12 + 18)/4
(07 + 13 + 18 + (Min + 18)/2)/4
(09 + 21 + Max + Min)/4
(02 + 06 + 10 + 14 + 18 + 22)/6
(0030 + 0330 + … 2130)/8
The *Köppen formula* is also sometimes used, as follows:
Mean $= n - k(n - \text{min})$, where n is the mean of three daily readings and k is seasonally dependent.

Each method gives a different mean. How the mean has been calculated at any one particular site is often not specified in the site's documents, and the method may also have changed with time without any record of the change. In most cases the daily maximum and minimum values or the spot readings are rarely quoted, just the means, and often only the monthly means are available, preventing any back-checking. Monthly means are simply the mean of the daily means. Annual means are the mean of the monthly means.

Time of taking the measurements

There is the important question of what time of the day the readings are taken. It has been traditional in the UK, for example, to make observations at 09:00, reading the 'present temperature' along with the readings stored by the maximum and minimum thermometers. Taking readings at this time of the day results in the minimum reading (which usually occurs in the early hours of the morning) applying to the same day as the readings are taken, while the maximum applies to the previous day (since the maximum usually occurs in mid-afternoon). The mean temperature, when readings are taken in the morning, is thus a mixture of two days' temperatures. However, in some countries the readings are made in the afternoon, and this places the maximum and minimum readings in the same day, resulting in a different (and perhaps more logical) mean. It is not vital to know which was used provided that the method was not changed. If it was changed, there will be a discontinuity in the record (see the next section). If there are good metadata, corrections can then be made.

Working in 'anomalies'

Given temperature readings from many stations, how can a mean temperature be calculated for an area? The first thought might be to take the mean of all the station readings in the area, but this would almost certainly produce a biased mean, the bias depending on where the stations are sited. If many of the stations are on untypically high ground the mean would be low; if the majority are in towns and cities, the mean would be higher. If most are clustered together in one region, the mean would be biased towards the temperature of that region.

There is also the problem of the spacing between stations. If they are not equally spaced on a grid (and few networks are), there will be a bias towards stations surrounded by a large area. Different 'weights' can be given to each site to allow for the size of the area they represent, such as the polygon method developed by Thiessen (1911) or the triangle method of Jones (1983), but the areas they represent may not be typical. There is no simple solution to this. We look at this problem again later.

In climatology, however, we are primarily interested in *changes* of temperature rather than *absolute* temperatures. (By 'absolute' is not meant temperature expressed in the absolute Kelvin scale; it just means the actual temperature as read off the thermometers and expressed in the usual everyday sense, such as 20.5°C.) For this reason, it is more convenient to work in temperature *differences* (temperature *anomalies*) rather than in the raw absolute temperature readings straight from the thermometers.

To calculate such anomalies, the raw temperatures (daily or monthly means) are subtracted from a temperature baseline, this usually being the average temperature over a 30-year period, the period chosen being the years when the measurements were at their most plentiful (or most reliable). The period varies, however, depending on the dataset being used, since different datasets often start and end at different times.

At present the reference period is from 1961 to 1990, but in the 1980s it was from 1946 to 1960. To calculate the anomalies at each individual station, an estimate is first made of the average for each individual station over the reference period for each day, month and year. (Very often the daily values are not available, in which case just monthly and annual means are calculated.) Each monthly mean from a station is then subtracted from its monthly 30-year mean, producing the anomaly (the amount by which this month's temperature differs from the long-term monthly mean). The same is done for annual anomalies. If the anomalies are now plotted, the year-by-year changes can be seen and a trendline can be added to smooth out the year-by-year variability to show the long-term trend.

Because many stations do not have complete records for the reference period, various methods have been developed to estimate the average by filling in the gaps using neighbouring stations.

There is, however, another important reason why anomalies are used instead of absolute temperatures, and that involves how anomalies from individual stations are merged to give areal means of temperature change. We will return to this in the section on 'Working in grids'.

Homogeneity

A question that must be asked before using any of the data collected by the meteorological stations around the world is 'are they free from error', in particular are they free from any errors that change over time? The word 'homogeneous' is used to describe data that are free from changes that result from non-climatic factors. Conrad and Pollak (1962) define a climate dataset as homogeneous as follows:

> A time series of monthly temperature averages is termed homogeneous if the variations exhibited by the series are solely the result of the vagaries of the weather and climate. Many factors can influence homogeneity (quoted in Jones et al. 1999, p. 174).

Indeed there are many factors that can affect the temperature measurements other than changes in the climate (or changes in the weather, for it is the day-by-day 'weather' that the stations record not the 'climate'; recall that the climate is the weather averaged over a long period, typically 30 years). These disrupting influences (explained fully in earlier chapters but briefly repeated here in summary) include:

- Stations being moved to a new site, often without a record of the move
- Changes in how the mean temperature is calculated, without any note of the changes
- Slow changes at the site itself, both rural and urban
- Poor maintenance of the screens, leading to increased heat-absorbing properties
- Slow changes in the surrounding countryside
- Undetected and uncorrected calibration changes of the thermometers.

So how can these influences be detected and removed? The verb 'to homogenise' is used to describe this procedure. Two methods have been developed, as described by Jones et al. (1986).

In the first approach, referred to as 'absolute homogeneity', *all* of the records used must be checked using station history records (metadata, see Appendix E), when available, as a guide to likely errors. The second technique, known as 'relative homogeneity', looks for inconsistencies between *neighbouring* stations that can then be used as a guide to *major* inhomogeneities in the dataset. The choice of which method to use depends on what the data are to be used for. If a single station record is to be used to follow local climate change, the first approach is the only way to proceed. For many other purposes the comparison between nearby stations is sufficient.

In the first method, one starts by assuming that all air temperature records contain errors (which is most likely), the procedure being to identify all potential errors from the station's recorded history and to apply corrections, station-by-station. But many stations have no recorded history, even in the most technologically developed and efficient countries, and even then corrections would not be perfect. Notes accompanying the Smithsonian Institution's and US Weather Bureau's *World Weather Records* (WWR) show that some of the data in the datasets have been corrected (or homogenised). The degree of testing, however, varies between countries, some being more diligent than others, and the records of the WWR include only scanty details of what was done. (I would always prefer to have the raw untouched data available in case it is necessary to investigate any problems – ideally both the raw and the corrected data would be available.) After 1950, when publication was taken over by the NOAA through the WMO, the amount of detail given was reduced further. At that time, however, the importance of corrections was less imperative than it is now, climate studies not then being a factor. In most cases, therefore, it is not possible to get back to the original raw data and so it would be impossible to check each of the thousands of stations used to detect climate change, and even more difficult to correct them. So 'Plan B' had to be adopted for mainstream climatology.

Since climatologists are primarily concerned with large areal means, the cruder second approach is seen as justifiable, even if much less meticulous. In practice, of course, this is fortunate, since it is, in reality, the only practical way to proceed. A more rigorous method would not be possible for most of the stations of the world because metadata are available from very few of them, especially over their full period of operation. In this simpler approach, differences in continuity and in trends between neighbouring stations are searched for. This technique assumes that within small areas the effects of climate change will be similar, and so the measurements should match each other. The sizes of the areas vary from 1,000 km^2

(32 × 32 km) to 100,000 km^2 (320 × 320 km). Assuming that trends will be the same across the larger areas is, of course, questionable, but this is not a precise method. Full details of how the comparisons were made are given by Jones et al. (1986). Briefly they are as follows.

For each station, the full length of its record is converted into annual anomalies using the monthly means (as described earlier). Any outliers are first removed. Each annual temperature anomaly is then compared with all of the other annual anomaly records from the same area and the difference between them plotted (Conrad and Pollak 1962). If an abrupt change is found (detected visually) a station-move or a change in observation time is suspected. By comparing a number of station-pairs, the faulty one can be isolated and the cause of the step-change found from the station metadata (if available). The record is then adjusted (or 'homogenised') so as to be compatible with the most recent part of the record. Such corrections are reliable enough for continental- or hemispheric-scale studies, but not for smaller-scale studies, although they are certainly better than the raw data.

The plots of the differences between station anomalies may also show a gradual change rather than a jump. This is more difficult to deal with. A sudden jump is fairly safely attributable to a site change or to a change in how the mean has been calculated. Changes in the instrument used can also cause a discontinuity. (When a change of instrument is made, it is important to have an overlapping period when both old and new are operated together for at least 12 months so as to be able to correct for any difference in performance.) Correcting trend differences is less straightforward, since the different trends might be real; the climate may not have changed in the same way at the two stations. In most such cases the record is flagged as uncorrectable, even though it may be correct. It is generally possible to pick out obvious urban heat island (UHI) effects, but other trends, including cooling trends, are excluded.

Now that we have each station record checked for homogeneity and expressed in anomalies, how can we get a mean temperature for a large area?

Working in grids

Different methods of gridding

The purpose of measuring temperature, and indeed all climate variables, is usually to estimate averages over an area. To get an average over anything other than a very small area, such as one field, a network of several instruments is necessary. Hydrology often requires, for example, the measurement of many variables across a river catchment in order to estimate its water resources and how the resources change with changing land use (Newson 1977).

Three methods have been developed to interpolate (insert) station data onto a regular grid, the most widely used being what is termed the *climate anomaly method* (CAM), which is the most common method, considered in detail below. The other methods are the *reference station method* (RSM) (Hansen and Lebedeff 1987) in which the world is divided into (about 80) approximately equal areas and a key (reference) station is identified within each area. The station must have a long record and be of good quality. Shorter records from within the area are then adjusted in a way that makes their averages equal to the composite of all the stations already incorporated. This allows all records to be used provided that they have an overlap of 10 years or more.

The *first difference method* (FDM) operates by taking the difference between each adjacent mean. For example, to get the first difference (FD) value for January 1981, the temperature for January 1980 is subtracted from it. The difference values for each time step (each year) are then averaged over all the stations in a region to form a regional FD series. If there are no gaps in the data, the result will be the same as conventional averaging (except that the time series will be shifted up or down so that the initial value is zero). A final series cannot be reconstructed from an FD series containing gaps unless the FD series is first combined with one or more other series. Full details of the mathematics and procedures are given by Peterson et al. (1998), who also made a comparison of all three methods and found that, using the same data as input, the differences between trends over the period 1880–1990 were only a few hundredths of a degree per 100 years. The differences were due more to the different methods of calculating the linear trend than to the differences between the methods (Vose et al. 2005). We now need to look more closely at the method that is most widely used – the CAM.

The climate anomaly method

As noted earlier, since climate studies are mostly concerned with temperature changes rather than absolute temperatures, anomalies are a convenient way of working. But another advantage of working in anomalies is that the technique assumes that all anomalies have the same significance, or 'weight', when calculating the mean anomaly over any area, up to the global, irrespective of:

- The altitude of the stations
- The time of the year, or season
- The climate in which the stations are operating
- The local microclimate.

The virtue of this method is that to obtain an average anomaly for an area, for a hemisphere or for the globe, the mean monthly anomalies in the area can simply be averaged. For now let us assume that this procedure is correct, for that is what everyone working with observations currently does (in one way or another), and it is on this basis that all present global and hemispheric temperature trends are based. In the next section, however, this assumption is questioned.

To calculate the average temperature anomaly for an area, it has become traditional to divide the surface of the globe into 'grids' of 5° latitude × 5° longitude and to average the anomalies of the stations located within the grids to get the areal mean. In the 1980s, 5° × 10° was used and more recently 1° × 1° has been added. Such grids have the disadvantage that they are not square but tapered, getting smaller with increasing latitude as the latitudinal lines get closer together. This can, however, be allowed for by weighting the grids proportional to their area. A constant-area method, as used in the reference station method, might be better. I have always visualised an alternative method in which the globe is divided into five- and six-sided polygons like a football. It would be more rational, although it would not change the result.

'Gridding' was initially performed by Jones et al. (1982) by selecting the six nearest stations to the grid's centre point. If there were no stations within 300 nautical miles of this point, a value was not calculated and the grid was flagged as missing – for that particular month. (Remember that the network is not static and there are gaps in the data, short and long, varying with time.) If all six stations were in the same quadrant (or corner) of the grid, again no value was calculated. When there were six stations in at least two quadrants of the grid, the mean overall anomaly was calculated using an inverse distance weighting; that is, the anomalies for each station were weighted according to the distance of the station from the centre of the grid, the more distant stations contributing proportionally less to the areal mean than the nearer stations. For stations closer than 30 nautical miles, the full value was used, with no weighting. For those wishing to follow the maths used in gridding, see Jones et al. (1982).

However, four years later, Jones et al. (1986) revised the gridding procedure so that each station was first associated with its nearest grid, and for each grid point all the available station anomalies were averaged, still using inverse distance weighting. This enabled all available stations to be used and also made it easier to cope with the coming and going of stations over time. Today's practice is to calculate the unweighted average of all of the stations within each box. P.D. Jones (personal communication, 2008) notes that changing the method of combining the data does not make much difference to the result.

Are anomalies weightless?

This section is an aside, reflecting on some questions about anomalies.

The technique of using anomalies to counteract or avoid or circumnavigate the difficulties caused by sites having different characteristics, different climates, varying altitudes and different seasons, assumes that while absolute temperatures are strongly affected by these factors (which they are), anomalies are completely immune from any of them. If we work in anomalies, it is reasoned, all these problems go away. Current practice, therefore, assumes that anomalies are all of equal 'weight' (influence or significance) wherever they are from on the planet and whichever season they were measured in. But is this correct?

An anomaly of x°C at 2,000 feet (approx. 600 m) altitude does not necessarily have the same 'weight' as an x°C anomaly at sea level nearby. An anomaly of x°C in a field may not have equal weight to the same anomaly in a nearby town. This is because temperature anomalies are not occurring in isolation from everything else. We need to look closer.

A monthly anomaly of, say, 0.2°C on the Greenland ice fields should not be given the same weight as an anomaly of 0.2°C in the Sahara desert or the Amazon rainforest (in the same month), because even though they are in the same hemisphere and measured in the same season, there are many different factors in operation, one such being the relative humidity (RH) of the atmosphere. The amount of water vapour in the atmosphere will affect the ratio into which the incoming net radiation is apportioned between the latent and sensible heat fluxes (see Chapter 1). In a humid atmosphere, there will be less evaporation than in a dry atmosphere (for the same incoming radiation), with more of the radiation being converted to sensible heat rather than to latent heat. That is, the available radiation will warm the air more than it would in a dry atmosphere (where evaporation could be greater), producing a greater temperature change or anomaly. In a dry atmosphere, on the other hand, there is the potential for more of the incoming energy to evaporate water (provided there is water available) and so the temperature anomaly should be given more weight when averaging it with other anomalies. This is not done at present – all anomalies being given the same weight whether they are from the Arctic in winter or the Sahara in summer.

Continuing on with the question of water vapour: at places where liquid water is plentiful, more of the incoming energy is likely to be converted into the latent heat flux (provided that the RH is not already high) because of greater evaporation, leaving less energy to warm the air, thus making the same temperature anomaly more significant than in a dry environment such as a desert or town (where evaporation is limited by the lack of free water). Water may also be available as soil moisture as well as open water. So the amount of water available should also be included when arriving at a weight for an anomaly

Another example: an anomaly of 0.2°C, at a specific site, when radiation is low (as in winter at high latitudes) should be given more weight than the same anomaly occurring at the same site under conditions of high radiation (summer), for more radiation has the *potential* to produce greater anomalies (even though it may not do so in practice, for entirely different reasons). An anomaly of 0.2°C in the summer should thus be given less weight than an anomaly of 0.2°C in the winter because there is more energy available in the summer and so the two cannot simply be averaged to produce an annual anomaly. And to these considerations of radiation amounts must be added the previously discussed questions of humidity levels and water availability, which will all be acting together.

Moreover, if we merge the Northern and Southern Hemisphere anomalies for January, February and March, we are mixing a northern winter with a southern summer. Although it is possible to average the anomalies for three months, north and south, I am uncertain how much real meaning this has. One global annual figure is even less helpful and more misleading, and probably inaccurate because of the non-weighting of the anomalies. All one can meaningfully do is look at changes in regions of similar climate, in the same hemisphere, with the same geography, in spring, summer, autumn or winter – not all of these things at once.

We must ask, therefore, whether anomalies have equal value everywhere, at all times, independent of everything else. The importance of an anomaly, or the weight it should be given when merging with other anomalies to produce an areal mean, depends on the level of radiation, on the humidity and on the amount of available water to be evaporated. Even wind speed will influence the amount of evaporation occurring and so should influence the weight given to the anomaly; for more wind will increase the amount of evaporation and so reduce the rise of temperature. And all of these factors act in concert, to different degrees.

Looking just at temperature changes in isolation is, therefore, an oversimplification. To make allowances for the various other influences would, of course, require measurements of variables other than temperature to be available, so that the temperature changes could be weighted correctly. This has not been done, in large part, I suspect, because it would make the calculations extremely complex and the data requirements too demanding. For instance, it would require a measure of humidity to be available along with the temperature. Willett et al. (2008) describe recent work, however, that has produced a global dataset (HadCRUH) quantifying the amount of water vapour in the atmosphere at the surface, and this goes part of the way to making more data available than just temperature. A measure of radiation and of how much water is available would also be necessary, the former often only recorded as daily sunshine duration. The amount of water would have to be estimated. Measurements of such variables are not readily available from the past, but could be estimated roughly. It is necessary that these matters be addressed, since they make the present estimates of temperature trends questionable, to a degree that it is very difficult to quantify. In the future, with new instrumentation, it would be possible to have all the relevant variables recorded at each station.

Merging gridded temperature anomalies to encompass complete hemispheres and seasons into one global figure has another problem, beyond that of anomaly weighting, for it obscures all the details. Detail is necessary if we are to understand and explain the intricacies of climate change, for only by looking at local areas and seasons will the complex mechanisms be detected. One annual figure masks all of this. This problem is expressed well by Jones et al. (1982, p. 110) when they say:

> Although a single hemispheric value is conceptually simple, it can obscure important spatial detail and may well obscure much of the physics associated with the majority of the observed variance.

Barnett (1978) comments likewise. Also, due to the Earth's elliptical orbit (see *Changes in the solar constant*, Chapter 1), the Northern Hemisphere receives more solar energy in its winter months than the Southern Hemisphere in its winters, while the Northern Hemisphere summers receive proportionally less than the summers in the Southern Hemisphere. This does not change in the short term (centuries), and it is changes that concern climatologists most, but it does mean that the seasons are not directly comparable. Global means have limited scientific use, even if the anomalies were weighted as suggested above.

Precipitation anomalies: a comparison It is worth reflecting on the similarities and differences between temperature and precipitation. Rainfall is quite a different kind of variable from temperature, being a quantity, an actual physical amount of a substance. Temperature is a state or condition, not an amount. So it is legitimate to take the measured amount of water caught by a rain gauge at one point and to sum it over any length of time from a minute to a century. This has real meaning; the water could be collected physically in a bucket. It is also valid to estimate the average fall over an area, different weights being given to each rain gauge to allow for the size of the area they represent, such as the polygon method developed by Thiessen (1911), or the triangle method of Jones (1983) (using the same principle as mentioned earlier for temperature). But this does not take account of whether the gauges, within their enclosing polygons, are representative of the area. If they are, there is no problem with stating that the annual rainfall for a given area is so many millimetres. The amount of rain landing at any one point is not affected by any physical processes occurring at the surface, so there is nothing to allow for, no weight to be assigned. (If they are in unrepresentative places, however, the mean will be biased.)

So while it is entirely meaningful, given well-sited stations, to state that the average amount of rain that fell on any particular area in one year is x millimetres, whether an *anomaly* of y millimetres of rain in the Sahara is equal in significance to the same anomaly in the Amazon or Antarctica, when calculating a global mean precipitation anomaly, is doubtful because they would each have quite different causes and connotations and so would need weighting accordingly. This is not done.

Nevertheless, since all climate-change analyses so far have been founded on the assumption that temperature anomalies *are* immune to all other influences, and so *are* absolutely equal and so *can* be averaged without any weighting whatsoever, we must now return to this line of thought. This section is a detour into unknown country, a question that I do not know the answer to, but a question which I have thought about often during my daily keep-fit walks by the Thames. It goes so much against current assumptions and procedures that we must leave it behind and return to the present status quo.

'Spatial degrees of freedom'

How many stations are needed to get an accurate estimate of the average temperature within an area? This is the same question as how many rain gauges are required to get a certain level of accuracy of rainfall measurement within a catchment.

Imagine a catchment completely covered in rain gauges with square funnels, all touching, with no space between. The area's average rainfall, for any period, would be the average reading of all the gauges. In practice, of course, most of the funnels are missing, just a few gauges being randomly scattered across the catchment, so how do you then calculate the mean rainfall? If the rainfall was always absolutely even across the catchment, only one rain gauge would be needed. But in practice precipitation varies greatly across space and time, especially in the case of convective rain, such as thunderstorms and in the tropics. So on a day-to-day basis there could be great variability in catch between gauges at different points in the catchment. Some might catch 20 mm while others, not far away, might catch nothing. With frontal rain the differences would, of course, be less, because the rain is more evenly spread in time and space. But even with convective rain, over a month the gauges will tend to agree more closely, while over years the differences will become increasingly less. So the answer to the question of how many rain gauges are needed over a catchment is that it depends mostly on the length of the interval of interest, on the absolute accuracy required and on the predominant type of rain; for daily intervals a great many more gauges are required than for monthly or yearly periods, to allow for the spatial variability of rainfall. Even over longer time intervals, however, there will be differences in catch due to local topographic variations and because higher ground tends to experience more precipitation, through the orographic effect of hills and mountains. So if the area is a flat plain, fewer gauges will be needed than in mountainous country. As noted earlier, the methods developed by Thiessen and Jones go part of the way to allowing for any spatially uneven distribution of rain gauges, but they cannot correct for variations due to the siting of gauges in non-typical microclimates.

Returning now to temperature, which is less spatially variable than rainfall – you can't have no temperature in one place and a lot of it in another – temperature is more 'catholic' than precipitation. If the whole area was covered with screens, say one every 10 m, the mean temperature of the area would be the mean of all their readings – as with the area completely covered with rain gauges. There would be enough of them to allow for all the different altitudes and different microclimates. But with just a few screens, the result obtained simply by combining their readings would not give an accurate mean, the precision depending on the exact positioning of the stations and the microclimates in which the stations were operating. If most stations were at high altitude, for example, the result would be biased cold, while if most were in towns it would be biased warm. How can we allow for the effects of different microclimates, as well as the uneven distribution of stations in most real networks, when calculating the mean temperature of an area? How close can we get to the ideal of the whole area covered in screens when there are just a few screens in practice? As with rainfall, the answer depends in part on the time span we are interested in, a different number of stations being needed for daily, monthly, annual, decadal and century-long periods. It also depends on the variability of the temperature within the region. In the UK, for example, it can vary greatly from day to day, while in equatorial climates it is fairly constant all year round.

This complex problem is dealt with by the statistical techniques of *spatial degrees of freedom* and the related topics of *spatial dependence, spatial analysis* and *analysis of variance* which are used to investigate the degree of 'associative dependence' (similarity) between independently measured values at different geographical points. In our case this involves asking how much agreement there is between temperature measurements made at different places, and how many stations are needed to get an 'accurate' answer. For readers interested in the mathematics of spatial analysis, an Internet search with any of the above terms will give more than enough background, for example:
http://en.wikipedia.org/wiki/Spatial_dependence
http://en.wikipedia.org/wiki/Spatial_analysis.

The latter site explains that, in a geographical context, nearby things usually share more similarities than those far apart. This is known as 'Tobler's first law of geography' and can be summed up as 'Everything is related to everything else, but nearby things are more closely related than they are to distant ones.'

This expresses simply and eloquently, in Orwellian style, how conditions in different parts of a desert or of a forest are likely to be similar, while a desert and a forest will be very different from each other.

Effects of missing grids

The same question arises at the larger scale. Because of the uneven distribution of stations around the globe, there are many grids where there are no stations for a mean anomaly to be derived. (We are talking here about the land surfaces; a similar situation arises over the sea.) Furthermore, not only is cover patchy at any one time, but cover changes over time as stations come and go.

The question of how many thermometer screens are needed to obtain an acceptable degree of accuracy over relatively small distances, dealt with in the previous section, can be extended to deal with the question of how the number of grids with data affects the accuracy of the global annual mean anomaly. To evaluate the effect on large-scale geographical means, Jones et al. (1982) investigated the effect of incomplete cover by comparing three different griddings:

An Optimum grid	(using all acceptable data in 1982)
A Reduced grid	(using only stations available in 1900)
A Maximum grid	(using a less stringent selection to include more stations).

This was done for the Northern Hemisphere, and from this it was concluded that the effects of incomplete and changing cover make little significant difference to the *annual* mean (the effect being comparable to the differences arising from the different datasets and methodologies used by different researchers). However, this only applies over a year or more, not over shorter time spans, when more grids containing data are needed. These investigations were repeated and extended to include the Southern Hemisphere (Jones et al. 1986; Jones 1988, 1994). From these we again see that the question centres on the time interval – the longer the interval of interest, the fewer data points (grids) are required.

It is perhaps not entirely surprising that there is a fair degree of agreement (over a year) between the three different griddings because they include many of the same stations, simply arranged in different combinations and numbers.

These findings, nonetheless, lead to the interesting conclusion that if all we want are global temperature means, as few as 100 or so stations are probably sufficient for each hemisphere. But, as stressed earlier, while global means may be what the lay public, media and politicians want, they obscure all the important detail and are of little use in advancing our understanding of climate-change processes; indeed they may inhibit insight by hiding the detail. It is in the local (regional) and short-term (seasonal) changes that the clues to the complexities (and so explanations) of climate change lie, not in one gross global figure. These matters will be explored further in the next chapter (Chapter 9), looking at the actual temperature trends around the globe, and also in the final chapter (Chapter 11), dealing with the planning of a new instrument network.

Datasets from the land

Just a few organisations have undertaken the huge task of assembling global temperature anomaly datasets, the most well known and comprehensive being from the Met Office's Hadley Centre, the University of East Anglia's Climatic Research Unit (CRU), the NOAA's National Climatic Data Center (NCDC), the NASA Goddard Institute for Space Studies (GISS) (see Parker et al. 1995a) and from Lugina et al. (2005). These are discussed, together with the SST datasets, in the final section of this chapter and in the next chapter.

Data from the oceans

All of the basics, such as homogeneity, anomalies, grids and spatial degrees of freedom are similar with regard to the land and the oceans and need not be repeated again here. Since the oceans cover 71% of the globe they are, area-for-area, climatologically more significant than the land and so will have proportionally more influence on the overall climate. Also we can probably learn more from ocean temperatures than from those over the land because SSTs are not distorted by the many different surfaces, land types, land uses, human settlements and mountains, which affect the weather over the land in ways difficult to quantify, and which make it hard to differentiate small climatic changes from the noise of inter-annual and inter-decadal variability (sophisticated statistics notwithstanding). Ocean temperatures also vary less both in time and space compared with the air temperature over the land.

Although many of the statistical basics are similar over land and sea, the actual measurement of temperature over the oceans is nevertheless done very differently from that over the land, as demonstrated in Chapter 5. As a brief reminder: air temperature over the sea (marine air temperature; MAT) is difficult to measure on ships because of the heating of the ship by the Sun, and in most cases it is replaced by the sea surface temperature (SST); nor is air temperature measured on most of the drifting buoys. Night marine air temperature (NMAT) is sometimes used, being unaffected by solar heating of the ship, but it only covers half of the daily cycle and so cannot give maximum and minimum values, making it impossible to calculate daily means. In any case, on a moving ship the mean would not refer to a particular location but to a large area.

The various methods of measuring SST, and their shortcomings, are fully explored in Chapter 5. In whatever way the SST is measured, however, the majority of observations come from moving platforms – ships and drifting buoys. This prevents long-term records of temperatures being made at fixed points, except from the few moored buoys that are in operation, and from a handful of islands.

In consequence, SST is not a daily mean of maximum and minimum readings at one point, as on land, but is an instantaneous reading at one point at one time, the large thermal inertia of the sea helping, however, to give an equivalent to an averaged-out reading. In place of the 30-year temperature mean generated for land temperatures at individual stations for the production of anomalies, a similar benchmark is derived for each 1° square of ocean for each day, each five-day period (pentad) and each year, based on all the readings made in that square over the 30 years from 1961 to 1990. Each reading made in that square is then subtracted from the 30-year mean to produce a temperature anomaly, just as in the case of the land temperatures.

Since each observation in any 1° grid is made by a different instrument each time, there is no equivalent to a bias or trend due to slow thermometer calibration drift. Since there is no 'screen' in which to deploy the thermometers, there is no screen to deteriorate, although there are many problems with buckets.

Weighting SST anomalies

The situation is simpler over the oceans because there are no altitudinal differences or different land types and urban areas, and there is always an unlimited amount of water available (except where the sea is frozen). However, there are still variations in humidity, in radiation, in sea surface condition and in wind speeds; all of which influence the division of the incoming radiation into the sensible and latent heat fluxes, just as on land. For this reason, a direct merging of all the anomalies from all of the oceans, from all seasons, latitudes and hemispheres, to give a hemispheric mean annual SST anomaly, without some form of weighting is, as on land, difficult to justify. It would also be hard to adjust the weightings because measurements of humidity and radiation are not available from ships or buoys to allow anomalies to be given weights, although sea-state and winds have been noted in ships' logs for centuries.

However, the errors are possibly slightly less than on land because the environmental differences are smaller, one area of the sea being very much like another over large expanses of ocean. Thus while it would perhaps seem reasonable to merge anomalies of SST over larger areas than over land, without the need to weight them, to extend this to the hemispheric or global scale, and to year-long time intervals, would seem less justifiable; is an SST anomaly of 0.2°C at the equator equivalent to the same anomaly near Antarctica in winter, the conditions being very different indeed?

We also need to take account of the great influence that the periodic swings between El Niño and La Niña, across much of the equatorial Pacific, have on SST; for example, they influenced the 1997/8 global temperature greatly. While there are no mountains and no cities at sea, there are more than enough other major differences to make the seas perhaps not as homogeneous as they might seem at first glance. For these reasons we need to look for clues about climate change over areas of similar oceanic conditions, and in looking at changes over long periods, to do this season-by-season rather than using a single blanket, global annual figure, which is as questionable over the oceans as it is over land.

Producing gridded values of SST

The raw SST data are 'aggregated' (combined) in a different way from air temperature on land. Each SST (or NMAT) value is accompanied by a location. Monthly fields of 1° square grids for each pentad and day of the year are calculated using all the data available from 1961 to 1990. Because many areas have no *in situ* data, a background climatology based partly on satellite data (adjusted to be in agreement with the *in situ* data) is used to fill the data-voids in each month from 1961 to 1990 (Parker et al. 1995b). From the 1° anomalies, 5° × 5° grid box anomalies are then produced for each month by averaging all of the 1° anomalies for each of the pentads that make up the month, thus making them compatible with the land grid sizes.

Datasets from the sea

The new International Comprehensive Ocean–Atmosphere Data Set (ICOADS) has been produced by blending the former COADS with the UK's Marine Data Bank and newly digitised data, including the US Maury Collection and Japan's Kobe Collection. All of these include data from buckets, ERI and drifting buoys, all adjusted to correct for their different biases. One of the principal datasets that has resulted from this combination is HadSST2, produced by the Hadley Centre of the Met Office (Diaz et al. 2002; Rayner et al. 2006). This supersedes its predecessor, MOHSST6, being based on an enlarged and improved set of observations (Rayner et al. 2006). This dataset provides the same information for the oceans as CRUTEM3 does for the land. It can be produced on a grid of any resolution.

Combining sea and land datasets

Equivalence between land air and sea surface data

As explained in Chapter 5, the use of sea surface temperature (SST) as a proxy for marine air temperature (MAT) requires that there be substantial agreement between the two, despite their very different natures and means of measurement. Investigations have shown that SST anomalies are approximately equal to those of the air about 10 m above the ocean surface (Jones et al. 1999). This is only true, however, on seasonal and longer time scales and in the open ocean (Rayner et al. 2003). Some regional discrepancies in trends have, however, recently been found even in the open ocean in the tropics (Christy et al. 2001), although these are quite small. This is because, as the atmospheric circulation changes, the relationship between SST and surface air temperature anomalies can change along with surface fluxes. So there is still some uncertainty about the strength of the link between SST and MAT. I am unhappy with the assumption that SST and MAT can be viewed as interchangeable, because they are very different in nature and because the equivalence only applies over seasons or years not over shorter times, and insufficient is as yet known about the relationship. This might be useful for annual means but, as noted elsewhere, the clues to climate change lie in the detail of seasonal and regional activity. But, as with much else in climate investigations, we have no other option but to use SSTs because nothing else is available.

Choice between SST and MAT

Due to the higher day-to-day stability of SST compared with MAT, and its lesser variability over distance, SST observations are considered to be more reliable, with fewer being required than those for MAT to achieve the same result (Trenberth et al. 1992). This would be the case even if the MAT observations had been as reliable as those of SST, which they are not. The better SST autocorrelation is because of the high thermal inertia of the water – it simply cannot change temperature as quickly as the air does – and because one part of the sea is very much like another over reasonable distances. However, from the point of view of equivalence, MAT would be a better choice when producing a global dataset because it is the same variable measured in the same way as air temperature over the land (land air temperature; LAT), with the same speed of response.

The blending process

The SSTs and LATs only need to be combined where land and sea coexist in the same grid box; otherwise they are presented entirely separately as grid boxes containing either LAT anomalies or SST anomalies, the assumption being (as above) that they are equivalent.

Where land and sea exist in the same grid boxes (around coasts and islands) they are merged to give a compound value of the temperature anomaly, this being weighted in such a way as to take account of the proportions of land and sea in the box. But because the land component from islands is potentially more reliable than the surrounding SST values, the land fraction is given at least 25% weighting (but no more than 75%), to prevent either being swamped by the other (Parker et al. 1994, 1995a). These constraints are needed because some grid boxes, which are nearly all sea, also contain one reliable land station on a small island, while other boxes are almost all land but contain many marine observations.

Recently Brohan et al. (2006) developed a new technique in which the more reliable value is given a higher weighting than the less reliable, reliability being evaluated by estimating the uncertainty of the data. How uncertainties are estimated is complex and those wishing to go into the maths should read Brohan et al.'s paper (which is well worth reading in its entirety as a very good summary of the present position).

Producing global datasets

A number of datasets have been produced: those that will be used in the next chapter to enquire into how temperatures have changed are the Climatic Research Unit's CRUTEM3 (previous version was CRUTEM2) for the land, and the Hadley Centre's HadSST2 for the sea. The merged dataset HadCRUT3 (formerly HadCRUT2) represents the two datasets, combining land and sea (Jones and Moberg 2003; Brohan et al. 2006). They can be downloaded from www.cru.uea.ac.uk/cru/data/temperature/.

In addition there are the datasets produced by the National Climatic Data Center (NCDC) (Smith and Reynolds 2005), NASA's Goddard Institute for Space Studies (GISS) (Hansen et al. 2001) and that produced by Lugina et al. (2005) – a Russian dataset.

The four differ mainly in how their global averages have been compiled, that for CRUTEM3 being weighted according to the proportions (0.68 × NH + 0.32 × SH), to allow for the different areas of land in the two hemispheres. The HadSST2 global dataset is weighted to account for the different areas of ocean in the two hemispheres, being in the proportion (0.44 × NH + 0.56 × SH). In the case of global datasets in which land and sea are combined, no weighting is necessary, since the whole area of the hemisphere is included.

The NCDC dataset is an area-weighted average of all available grid-box anomalies worldwide. The GISS dataset is the average of the anomalies for the latitudinal zones 90°N to 23.6°N, 23.6°N to 23.6°S and 23.6°S to 90°S, with weightings of 0.3, 0.4 and 0.3, respectively, proportional to their total areas. The Lugina dataset is weighted (NH + 0.866 × SH) / 1.866, with latitudes south of 60°S being excluded because there are very few data from there. As a result of these different ways of combining the grid-box anomalies, the recent global trends are largest in

CRUTEM3 and NCDC, because they give more weight to the NH land, where recent upward temperature trends have been greatest.

The treatment of gaps in the data gives rise to further differences. The GISS gridding method, for example, favours isolated island and coastal sites, which reduce recent trends, while Lugina et al. (2005) also reduce recent trends owing to their optimal interpolation method that tends to adjust anomalies downwards where there are few observations nearby (Hurrell and Trenberth 1999). The NCDC analysis is higher than CRUTEM3 by between 0.1°C and 0.2°C in the first half of the twentieth century and since the late 1990s, probably because its anomalies have been interpolated to fill in any gaps (Vose et al. 2005). Differences also arise because the number of stations used by CRUTEM3, NCDC and GISS differs (4,349, 7,230 and >7,200, respectively), although many of the basic station data are held in common. CRUTEM3 station numbers are lower because their method requires there to be sufficient data between 1961 and 1990 to allow the calculation of anomalies (Brohan et al. 2006). Further differences may have arisen from differing homogeneity adjustments. By the time of publication there may have been some slight changes (for example there may have been a small adjustment to the SST just after 1945 to allow for a change in how SST was measured during the Second World War; see Chapter 9).

We next look at what these datasets tell us about how the temperature has changed over the last 160 or so years.

References

Barnett, T. P. (1978) Estimating variability of surface air temperature in the Northern Hemisphere. *Monthly Weather Review* **106**, 1353–67.

Brohan, P., Kennedy, J. J., Harris, I., Tett, S. F. B. and Jones, P. D. (2006) Uncertainty estimates in regional and global observed temperature changes: a new dataset from 1850. *Journal of Geophysical Research* **111**, D12106.

Christy, R. D., Parker, D. E., Brown, S. J., Macadam, I., Stendel, M. and Norris, W. B. (2001) Differential trends in tropical sea surface and atmosphere temperatures since 1979. *Geophysical Research Letters* **28**, 183–6.

Conrad, V. and Pollak, L. D. (1962) *Methods in Climatology*. Harvard University Press: Cambridge, MA.

Diaz, H., Folland, C., Manabe, T., Parker, D., Reynolds, R. and Woodruff, S. (2002) Workshop on 'Advances in the use of historical marine climate data'. *WMO Bulletin* **51**, 377–80.

Hansen, J. E. and Lebedeff, S. (1987) Global trends of measured surface air temperature. *Journal of Geophysical Research* **92**, 13345–72.

Hansen, J., Ruedy, R., Sato, M., Imhoff, M., Lawrence, W., Easterling, D., Peterson, T. and Karl, T. (2001) A closer look at United States and global surface temperature change. *Journal of Geophysical Research* **106**, 23947–63.

Hurrell, J.W. and Trenberth, K.E. (1999) Global sea surface temperature analyses: multiple problems and their implications for climate analysis, modelling and reanalysis. *Bulletin of the American Meteorological Society* **80**, 2661–78.

Jones, P.D. (1988) Hemispheric surface air temperature variations: recent trends and an update to 1987. *Journal of Climate* **1**, 654–60.

Jones, P.D. (1994) Hemispheric surface air temperature variations: a reanalysis and an update to 1993. *Journal of Climate* **7**, 1794–802.

Jones, P.D. and Moberg, A. (2003) Hemispheric and large-scale surface air temperature variations: an extensive revision and update to 2001. *Journal of Climate* **16**, 206–23.

Jones, P.D., Wigley, T.M.L. and Kelly, P.M. (1982) Variations in surface air temperature: Part 1. Northern Hemisphere, 1881–1980. *Monthly Weather Review* **110**(2), 59–70.

Jones, P.D., Raper, S.C.B., Bradley, R.S., Diaz, H.F., Kelly, P.M. and Wigley, T.M.L. (1986) Northern Hemisphere surface air temperature variations: 1851–1984. *Journal of Applied Meteorology* **25**(2), 161–79.

Jones, P.D., New, M., Parker, D.E., Martin, S. and Rigor, I.G. (1999) Surface air temperature and its changes over the past 150 years. *Reviews of Geophysics* **37**(2), 173–99.

Jones, S.B. (1983) *The Estimation of Catchment Average Point Rainfalls*. IH Report No. 87. Institute of Hydrology: Wallingford, UK.

Lugina, K.M., Groisman, P.Ya., Vinnikov, K.Y., Koknaeva, V.V., Speranskaya, N.A. and Curtis, C. (2005) Monthly surface air temperature time series area-averaged over the 30-degree latitudinal belts of the globe, 1881–2004. In: *Trends: A Compendium of Data on Global Change*. Carbon Dioxide Information Analysis Center, Oak Ridge National Laboratory, US Department of Energy: Oak Ridge, TN. Available at http://cdiac.esd.ornl.gov/trends/temp/lugina/lugina.html.

Newson, M.D. (1977) *Selected Measurement Techniques in Use at Plynlimon Experimental Catchments*. IH Report No 43. Institute of Hydrology: Wallingford, UK.

Parker, D.E., Jones, P.D., Folland, C.K. and Bevan, A. (1994) Interdecadal changes of surface temperature since the late nineteenth century. *Journal of Geophysical Research* **99**, 14373–99.

Parker, D.E., Jackson, M. and Horton, E.B. (1995a) *The GISST2.2 Sea Surface Temperature and Sea-Ice Climatology*. Climate Research Technical Note 63. Hadley Centre, Meteorological Office: Bracknell, UK.

Parker, D.E., Folland, C.K. and Jackson, M. (1995b) Marine surface temperature: observed variations and data requirements. *Climatic Change* **31**, 559–600.

Peterson, T.C., Karl, T.R., Jamason, P.F., Knight, R. and Easterling, D.R. (1998) The first difference method: maximising station density for the calculation of long-term global temperature change. *Journal of Geophysical Research* **103**, 25967–74.

Rayner, N.A., Parker, D.E., Horton, E.B., Folland, C.K., Alexander, L.V., Rowell, D.P., Kent, E.C. and Kaplan, A. (2003) Global analyses of sea surface temperature, sea ice, and night marine air temperature since the late nineteenth century. *Journal of Geophysical Research* **108**(D14), 4407, 2-1–2-22.

Rayner, N.A., Brohan, P., Parker, D.E., Folland, C.K., Kennedy, J.J., Vanicek, M., Ansell, T.J. and Tett, S.F.B. (2006) Improved analyses of changes and uncertainties in sea surface temperature measured in situ since the mid-nineteenth century: the HadSST2 dataset. *Journal of Climate* **19**(3), 446–69.

Smith, T.M. and Reynolds, R.W. (2005) A global merged land and sea surface temperature reconstruction based on historical observations (1880–1997). *Journal of Climate* **18**, 2021–36.
Thiessen, A.H. (1911) Precipitation for large areas. *Monthly Weather Review* **39**, 1082–4.
Trenberth, K.E., Christy, J.R. and Hurrell, J.W. (1992) Monitoring global monthly mean surface temperatures. *Journal of Climate* **5**, 1405–23.
Vose, R.S., Wuertz, D., Peterson, T.C. and Jones, P.D. (2005) An intercomparison of surface air temperature analyses at the global, hemispheric and grid-box scale. *Geophysical Research Letters* **32**, L18718.
Willett, K.M., Jones, P.D., Gillett, N.P. and Thorne, P.W. (2008) Recent changes in surface humidity: development of the HadCRUH dataset. *Journal of Climate* **21**, 5364–83.

9

Changes in air and sea temperatures

The datasets

The datasets discussed at the end of the previous chapter – CRUTEM3, HadSST2 and HadCRUT3 – are used extensively in the Intergovernmental Panel on Climate Change's Fourth Assessment Report (AR4; IPCC 2007), along with those from the National Climatic Data Center (NCDC), NASA's Goddard Institute for Space Studies (GISS) and that produced by Lugina et al. (2005). As explained, they differ mainly in how their global anomalies have been compiled from the individual grids and hemispheric datasets, but because the end results are very similar (and to keep this chapter a reasonable length), we concentrate on the CRUTEM3, HadSST2 and HadCRUT3 datasets and see what they tell us about temperature changes since 1850.

The HAD/CRU data run from 1850 to 2008 in the form of monthly and annual anomalies (referred to a 1961–90 baseline) for each of the 5° × 5° grid boxes, as well as the grids combined to give datasets for the whole Northern Hemisphere (NH), the whole Southern Hemisphere (SH) and the whole globe for the land, for the sea and for both combined. For reasons of space, individual grid boxes cannot be displayed and discussed, since the amount of data is extremely high and they would require a large book to themselves. To fit into one chapter of one book, we have to limit the discussion to just the hemispheric and global datasets – despite my reservations (expressed earlier) about the validity of gross merging into hemispheres. The datasets used are as follows:

- Northern Hemisphere (NH) land air temperature
- Southern Hemisphere (SH) land air temperature
- Global (GL) land air temperature
- Northern Hemisphere (NH) sea surface temperature
- Southern Hemisphere (SH) sea surface temperature
- Global (GL) sea surface temperature

- Northern Hemisphere (NH) land and sea combined
- Southern Hemisphere (SH) land and sea combined
- Global (GL) land and sea combined.

These datasets are available in two forms: *unadjusted* and *variance-adjusted*. Variance adjustment is a statistical procedure used to reduce the effect of there being only a small number of observations in a grid box. The procedure does not introduce biases into the grid-box series but on larger space scales, a bias is introduced. Whether adjusted or unadjusted data should be used depends, therefore, on what the application is. If it is important that grid-box anomalies have a consistent variance, then variance-adjusted data are more suitable. Otherwise, it is better to use unadjusted data, in particular for global and regional scales; for this reason the unadjusted data are used here (Brohan et al. 2006).

The GISS data are available from 1880 to 2008 as anomalies (referred to a 1951–80 baseline), as individual grid datasets, as well as in the following merged forms:

- Global (GL) land for each month, each season (JFM, etc.) and year (J–D and D–N)
- Northern Hemisphere (NH) land for each month, each season (JFM, etc.) and year (J–D and D–N)
- Southern Hemisphere (SH) land for each month, each season (JFM, etc.) and year (J–D and D–N)
- Global (GL) land and sea for each month, each season (JFM, etc.) and year (J–D and D–N)
- Land divided into ten latitudinal zones (24N–24S, 24N–90N, 24S–90S, 64S–44S, 44S–24S, 24S–0, 0–24N, 24N–44N, 44N–64N, 64N–90N)
- Land and sea divided into the same ten latitudinal zones.

However, lack of space does not allow these to be used in this book.

Trendlines

To show trends on graphs, it is useful to apply a low-pass filter to the data. Mann (2004) discusses smoothing time series and especially how to treat the ends. (If a sudden trend starts near the end of the sequence a trendline may not follow it correctly, since it is averaging over a period of years.) The IPCC and most climatologists use complex low-pass Gaussian filters, but binomial filters work just as well. I have used the ‘moving average’ method provided by Microsoft Excel software. However, it has one problem: if the averaging time is set at 10 years, the calculated average of this period is plotted at the end of the 10 years, whereas it would be better if it was plotted in the middle. This results in the trendline being shifted to the right by half the number of years averaged. To minimise this effect, an average of five years has been used, the offset from the centre years then being only three years, which is hardly noticeable (but should be kept in mind when looking at the graphs). An effect of this is that the trendline follows shorter time-interval changes and so becomes less smooth, although this has the advantage that the faster changes become more visible and any 'end effects' are minimised.

Causes of temperature change

Measurements versus model projections

Before looking at the graphs of temperature change, it is useful to review what might have caused the observed changes. The climate has always been in a state of flux, but we are not looking here at changes over thousands or millions of years (see Chapter 1) but at those over the instrumental record, since 1850, extending back in England to 1660 (see *Central England Temperature since 1659* later in this chapter).

The commonly held view is that temperature change can be summed up in one global graph and that much of the post-1975 temperature increase has an anthropogenic element. It must be noted, however, that any evidence for human influence on temperature change comes only from the models and not from the actual measurements. All that the measurements show is by how much the temperature changed. These changes could be due to many things, these being listed below. The question of model accuracy and reliability is not a topic that can be covered in this book, but in the technically detailed pages of the comprehensive 995-page IPCC AR4 (IPCC 2007) (as opposed to its much more widely read and less technical, 'Summary for Policymakers') things are not presented as being so certain. The second paragraph of the Technical Summary (p. 21) stresses the uncertainties in the following unambiguous terms:

> While this report provides new and important policy-relevant information on the scientific understanding of climate change, the complexity of the climate system and the multiple interactions that determine its behaviour impose limitations on our ability to understand fully the future course of Earth's global climate. There is still an incomplete physical understanding of many components of the climate system and their role in climate change. Key uncertainties include aspects of the roles played by clouds, the cryosphere, the oceans, land use and couplings between climate and the biogeochemical cycles.

The areas of uncertainty, therefore, cover quite a large spectrum. This important caveat is not sufficiently well aired publicly (and there are many more cautions like it in all the IPCC reports). Additionally, in Chapter 10 of the IPCC's AR4 the limitations of climate models are also discussed in great detail, too complex for anyone but an expert to follow. To paraphrase Donald Rumsfeld: 'There are things we know we don't know, but there also things we don't know we don't know'. This is not to deprecate models, but to remind readers that they have limitations.

The *known* causes of climate change are reviewed below, but there are also numerous complex interactions within the system, each affecting the others in a string of feedbacks at many scales, and it is questionable whether the climate is simple enough, or well enough understood yet, to be modelled reliably over long periods ahead. Only time will tell. Nevertheless, many of the year-to-year and decadal trends can be related to several important influences – changing ocean temperatures, atmospheric circulation changes and volcanic eruptions. In particular, there are what are termed *teleconnections*, which we need to take a look at.

Teleconnections

Although in many ways chaotic, the global circulation of the atmosphere has a number of patterns which it 'prefers' to be in, affecting the weather and hence the climate at the surface. These are known as *teleconnections* because weather conditions are related to each other over large distances. Understanding teleconnections – and how they produce droughts and floods, hot and cold spells – is central to understanding regional climate change. These interrelationships were discovered by the British meteorologist, Sir Gilbert Walker, in the late nineteenth century as he correlated records of atmospheric pressure, temperature and precipitation over large areas of the world (Walker and Bliss 1932).

The most familiar and powerful teleconnection, the *Southern Oscillation*, affects much of the tropical Pacific Ocean ('Oscillation' because the barometric pressures oscillate between two states, or between two preferred patterns). Because of its all-pervasive global influence, it is useful to look at the Southern Oscillation in some detail. It also demonstrates how climate at one point on the Earth's surface is tied and related to another, thousands of kilometres distant. The pressures do not switch in isolation from everything else, however. Because they are linked to sea surface temperatures, they affect each other in a complex and circular fashion, which we will now explore.

The Southern Oscillation

The Southern Oscillation concerns the links between sea-level barometric pressure at Tahiti with that at Darwin in Australia; first one being higher, then the other, with the swing from one state to the other being termed the Southern Oscillation. Switches occur annually or every few years. Teleconnections refer to large areas, but are usually characterised (for convenience and practicality) by barometric pressure measurements made at just a few specific localities, in this case at Tahiti and Darwin. To appreciate the causes of the pressure oscillations we must first look at the changes in the SST across the whole of the tropical Pacific.

El Niño and La Niña A bit of history first: every year, about Christmas-time, a warm southerly-flowing current, originating from around the Galapagos Islands, replaces the very cold northward current along the coasts of Peru and Ecuador. Every few years the southern current is much warmer (causing damaging disruptions to the sea life around the Galapagos Islands). This has been known for centuries by the local South American inhabitants, but it was only recently recognised that these periodic events extend over the entire tropical Pacific (Bjerknes 1969; Allen et al. 1996; Slingo 1998). While originally the term *El Niño* was used by fishermen to describe the local annual warm current (it means 'the Christ child' in Spanish, because of its time of onset) it is now used to describe the large-scale warmer events across the whole of the Pacific. More recently the term La Niña ('the girl child') was introduced (Philander 1990) to describe the opposite conditions.

As Figure 9.1 illustrates, the ocean is composed of two layers: a warm very shallow top layer about 100 m deep above a cold layer extending down to depths of 4,000 m or more, the two layers being separated by the *thermocline*, a thin layer across which the temperature gradient is high.

Under 'normal' conditions, moderate trade winds force the warm surface water of the Pacific westwards, causing the thermocline to rise towards the surface in the east, where it reduces the sea surface temperature compared with that in the west (Figure 9.1a). Due to the high surface temperature of the ocean in the west, convective storms occur in profusion over Indonesia and Australia, with heavy rainfall and thunder. The thermals rise to near the tropopause (Figure 9.2a), resulting in upper-level westerly winds that travel in a closed 'Walker circulation' as far as South America, where they subside, causing high barometric pressure and consequently drought, returning at the surface to the west as moderate trade winds.

When 'La Niña' conditions prevail, a more extreme form of the 'normal' state exists, in which the trade winds are more intense, forcing the warmer waters further west, raising the thermocline in the east still further, causing greater cooling of the sea surface (Figure 9.1a). The wind and precipitation patterns are retained but are intensified.

When some small perturbation in the trade winds causes the wind to lessen slightly, the warm water starts to move back eastwards and this enhances the reduction of the trade winds, which causes a still further eastward movement of the warm surface, combined with a sinking of the thermocline as it levels off to become almost horizontal (Figure 9.1b). 'El Niño' is now established, the warm sea producing large cumulus clouds with high rainfall over the central parts of the Pacific (Figure 9.2b).

There are now two Walker circulations, the much warmer central Pacific producing a shift of the precipitating cumulonimbus clouds eastwards from over

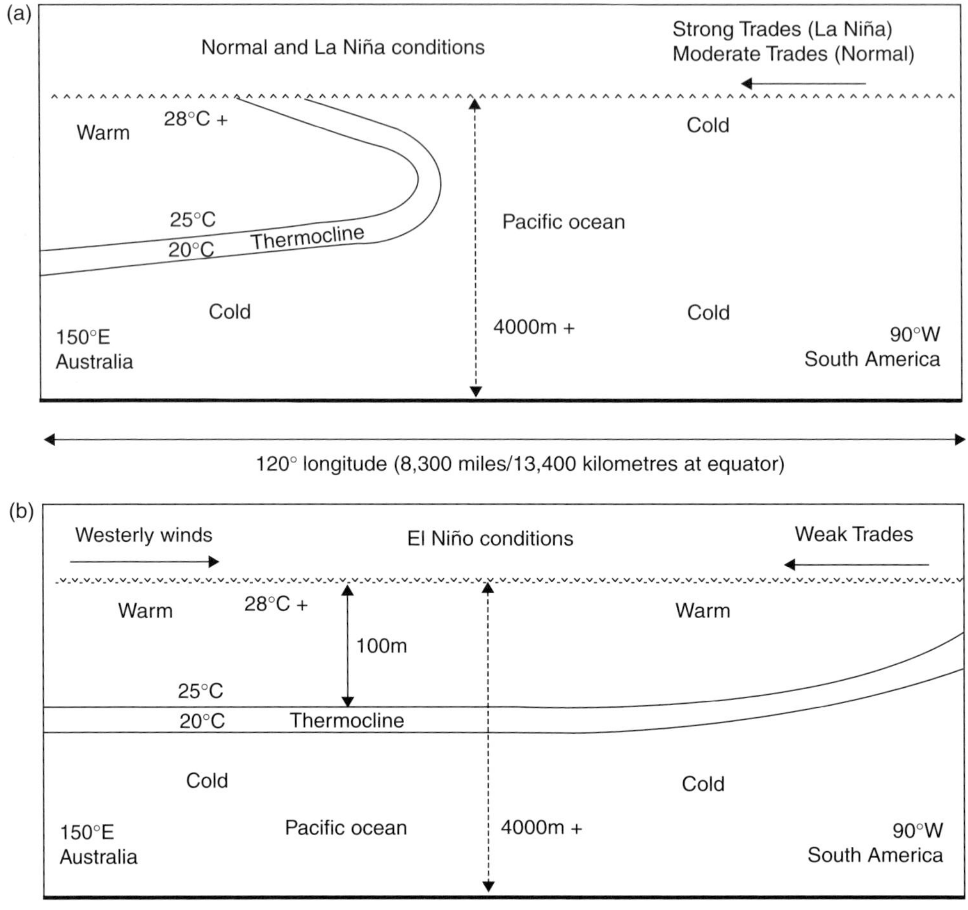

Figure 9.1 El Niño and La Niña are the names given to the two different states of the surface waters of the equatorial Pacific. (a) La Niña can be considered as a more extreme form of the 'normal' conditions, the strong trade winds holding the warmer surface water to the west allowing cooler water to rise to the surface in the east and central Pacific. (b) El Niño is the alternative state caused by the trade winds weakening, allowing warmer water to cover the entire Pacific. The cause of these alternating states and their effects are discussed in the text.

Australia, accompanied by low barometric pressure, with Tahiti being in the midst of this new rainfall activity, while the barometric pressure over Australia rises due to the descending cold air – the city of Darwin being near its centre. Over to the east, in South America, the descending Walker circulation produces a comparable high pressure. Sooner or later, however, the trades pick up again and force the warm water back to the west and La Niña returns.

What causes these changes in wind that result in the oscillation of barometric pressure? Although the changes in sea surface temperature (SST) cause the

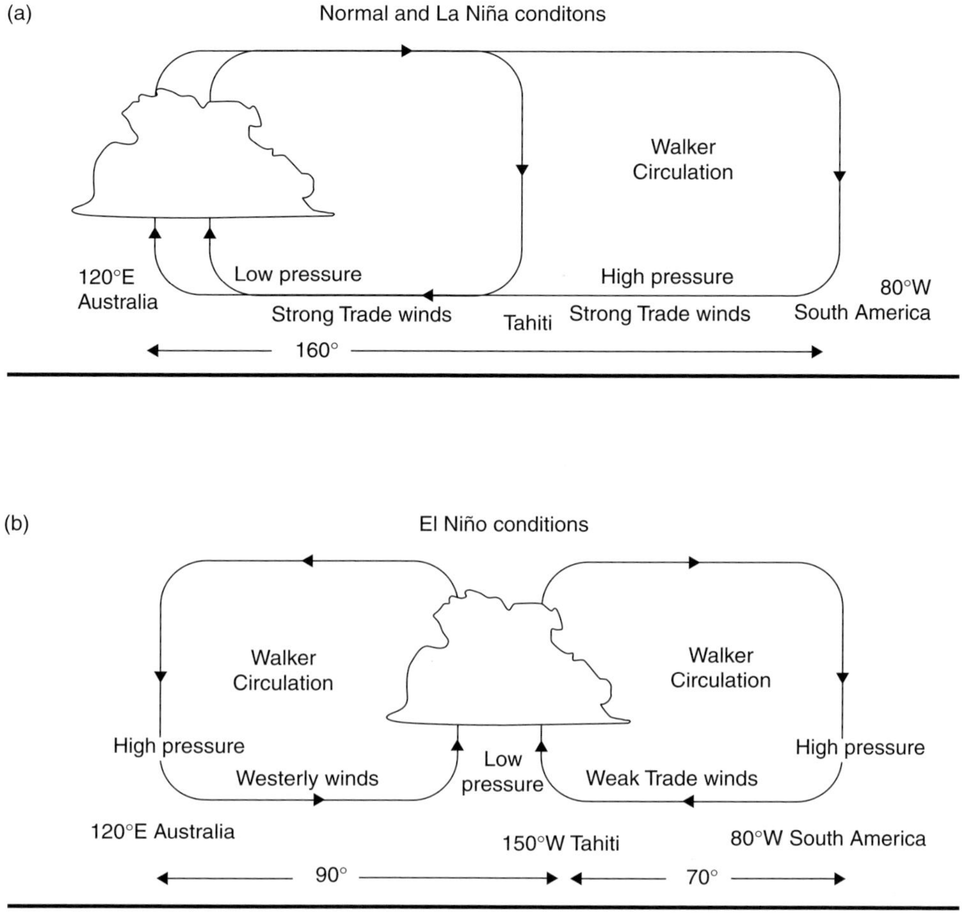

Figure 9.2 One of the effects of the changing sea surface temperatures brought about by the switch between El Niño and La Niña is to change the circulation patterns of the atmosphere over the Pacific as shown. (a) In the La Niña phase there is one single large Walker circulation producing higher pressure in the east, with lower pressure over Australia with resultant cumulus cloud formation and precipitation. (b) In the El Niño phase the convective activity occurs more over the central regions of the Pacific producing two Walker circulations with a resultant low pressure over Tahiti and higher pressure over Australia and South America. This figure should be looked at in combination with Figure 9.1, because the processes are interrelated.

changes in the atmosphere – in the form of wind, precipitation and barometric pressure modifications – the atmosphere also causes the changes in the SST. So which comes first?

The atmosphere and ocean are not equals in this and are not balanced, the atmosphere being lightweight, changing quickly in response to changes over the ocean in a matter of days or weeks. The ocean, on the other hand, having a high degree of inertia, takes months to adjust to wind changes. So the ocean state at any one time is not determined simply by the present winds, since it is still responding and adjusting to earlier winds. It has, in effect, a memory in the form of waves. The waves act below the surface, moving the thermocline up in one place and down in another, and this affects the transition from one phase to the next. This makes it important to measure the ocean conditions so as to be able to anticipate future changes (see Chapter 5).

Each phase typically lasts about 9–12 months, usually beginning to form during June–August, reaching a peak from December to April, when it has most effect on precipitation, and decaying away during May–July of the year following. Some episodes can, however, last up to two years (or sometimes longer), and their periodicity is quite irregular, although on average El Niño and La Niña occur every three to five years.

The combined changes of SST and barometric pressure, together, are known as the *El Niño Southern Oscillation* or ENSO.

The Southern Oscillation Index If the SST variations (as departures from 22°C) in the central and eastern Pacific are plotted, the cyclic changes can be seen to coincide with the plotted difference in pressure (Tahiti–Darwin). During an El Niño event, the pressure at Tahiti is less than at Darwin and so this is known as the negative phase. In this state, the SST is above normal. When the pressures reverse, during La Niña, the phase is said to be positive and the SST is below normal. To quantify this, the concept of the *Southern Oscillation Index* (SOI) was created (Figure 9.3a) (see Appendix F for details of the calculation). Because many of the trends in temperature are a result of variations in these oscillations, it is useful to look carefully at the SOI. It is also useful to keep a plot of the SOI (Figure 9.3a) to hand for comparison when looking at the temperature trends seen in Figures 9.6–9.12.

In Figure 9.3a, the dot-dashed line marks the coldest point in most temperature graphs, coincident with a persistent La Niña around 1910. This was followed around 1915 by a large and persistent El Niño (left of the dotted line) which raised temperatures briefly towards the 1961–90 mean. This was followed by a sudden switch to an even larger La Niña (to the right of the dotted line), taking temperatures down again. La Niña events then continued up to the late 1930s, slowly reducing in amplitude until the switch to another large El Niño event from 1939 to 1942 (to the left of the dashed line), bringing temperature anomalies back up to, and sometimes above, the 30-year mean in all of the graphs.

For the next 35 years, La Niña episodes dominated, with just a few small El Niño events interspersed, the period ending with a large La Niña (to the left of the solid line). At this point there occurred what is termed a 'climate shift', which

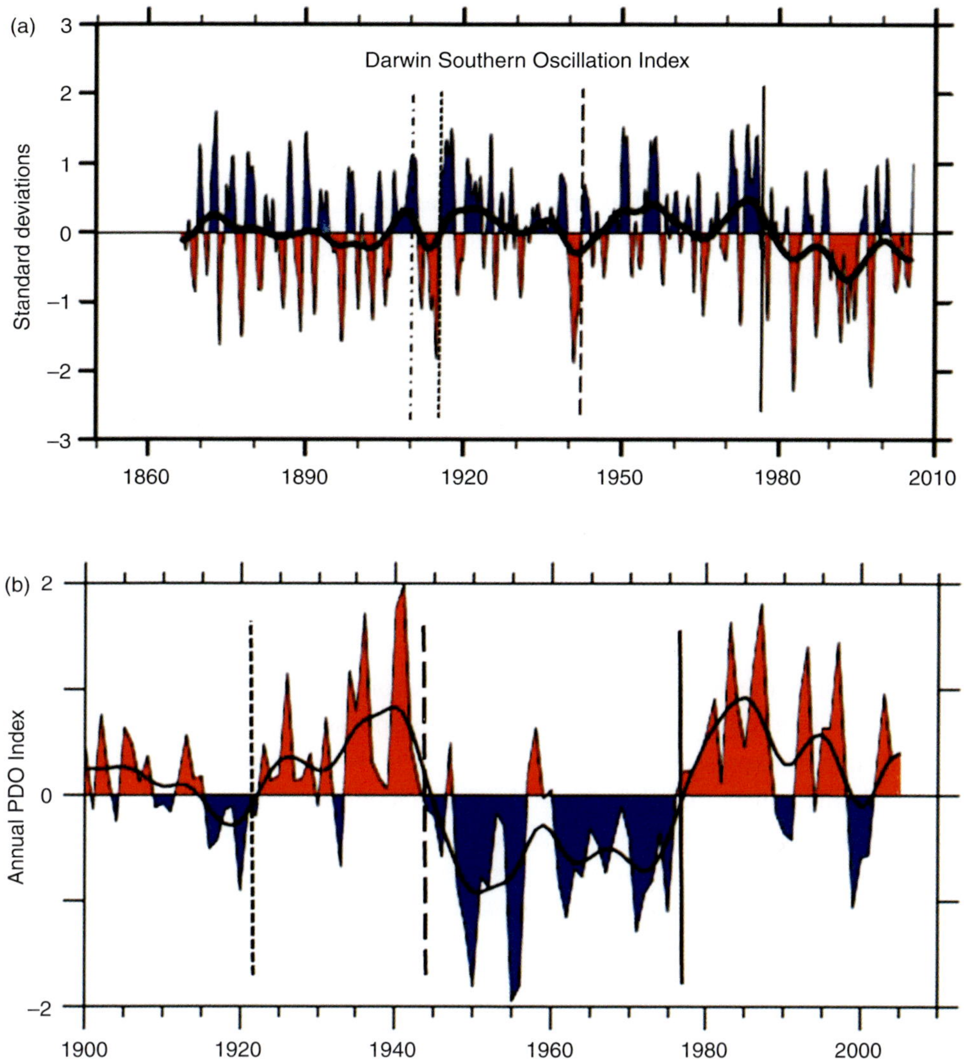

Figure 9.3 (a) The Darwin-based Southern Oscillation Index (SOI) in normalised units of standard deviation, from 1866 to 2005, showing monthly values, smoothed so as not to show fluctuations of less than eight months. The *black curve* shows decadal variations; *red* values indicate positive sea-level pressures at Darwin (El Niño and warmer surface waters); the *vertical lines* indicate where there are sudden changes in the index (at 1915 – *dotted line*, 1942 – *dashed line* and 1977 – *solid line*). See text for discussion (reproduced from the IPCC AR4 2007). (b) The Pacific Decadal Oscillation (PDO) index showing SSTs north of 20°N from 1901 to 2004. The *vertical lines* show the changes in the index (at 1922 – *dotted line*, 1944 – *dashed line* and 1977 – *solid line*) (graphs, but not the vertical lines, reproduced from the IPCC AR4 2007, updated from Mantua et al. 1997).

marked the beginning of the recent warming period. This shift was caused by a switch from dominating large and frequent cold La Niña events to dominating large, frequent and increasing warm El Niño events, with a few small La Niña events interspersed. It is this switch from La Niña to El Niño episodes that has been the cause of much of the recent warming, globally. The large 1998 El Niño, in particular, dominates all of the graphs. The last large El Niño was in 2005. More recently (not shown in Figure 9.3), over the last few years, there has been a return to larger La Niña events, causing some recent cooling.

There are around ten other well-studied 'oscillations', all with their own indexes (see IPCC, 2007, Box 3.4, p. 287), but we need not look at them all in the same detail as we have with the Southern Oscillation as the mechanisms by which teleconnections affect climate have been demonstrated sufficiently by this particular example. The other oscillations will only be covered briefly.

The Pacific Decadal Oscillation

The Pacific Decadal Oscillation (PDO) is another pattern of climate variability which switches phase, in this case every 20 to 30 years. The PDO is expressed as warm or cool surface waters in the Pacific north of 20°N. During a 'warm' (or positive) phase, the western Pacific becomes cool and part of the eastern ocean warms; during a 'cool' (or negative) phase, the opposite occurs. As with the SOI, the PDO can be expressed as an index. Changes in the PDO index (Figure 9.3b) are simpler, being longer term than the SOI (which changes every couple of years or so, although its trends are longer term). The changes in polarity or phase of the PDO index occur at almost the same time as the large long-term changes of the SOI.

To the left of the dotted line in Figure 9.3b, around 1918, the index reaches a low, switching to a warming phase that increases up to the dashed line in the mid-1940s, coinciding with the dashed line in the SOI graph. From then until the solid line, around 1977, a cold phase persists, switching to a warm phase in concert with the SOI, the two complementing each other. The PDO index and the SOI are important because they dominate the temperature trends seen in the graphs of temperature anomalies.

There is also an *Inter-decadal Pacific Oscillation* (IPO or ID) having similar SST and sea-level barometric pressure patterns, with a cycle of 15–30 years, affecting both the north and south Pacific. There is also the quasi-decadal oscillation (QDO) with a period of 8–12 years.

Other oscillations

Other oscillations have also been found, such as the *North Atlantic Oscillation* (NAO), which describes the variation in barometric pressures between Iceland and the Azores (or Lisbon or Gibraltar) and the related *North Annular Mode* (NAM; also known as the *Arctic Oscillation*), the *Southern Annular Mode* (SAM; previously known as the *Antarctic Oscillation*), the *Atlantic Multidecadal Oscillation* (AMO; a mode of natural variability of SSTs in the North Atlantic) and the *Indian Ocean Dipole* (IOD). All of these are quantified by 'indexes' that compare readings at two stations separated by large distances. Plots of these indexes show cyclic changes, similar to those illustrated in Figure 9.3.

Influence of the oscillations

The IPCC comment that these teleconnection patterns account for much of the seasonal to inter-annual variability and conclude that decadal variations in teleconnections complicate the interpretation of climate change considerably, noting that the ENSO and the PDO have varied throughout the past 100 years on an inter-decadal time scale. Of particular note are the systematic changes in the ENSO, with the enhanced El Niño activity since 1976, such that much of the recent increase in temperature is due to the large number of major El Niño events since the mid-1970s, through all of the 1990s and into the first half of the present decade. Since then they have reduced, and La Niña events have started to re-emerge, accounting for the recent fall in temperature anomalies. At the same time, the PDO has also undergone similar, synchronised, changes, combining with the ENSO changes.

The ocean surface temperatures are, in consequence, constantly changing as the teleconnections switch phase and interact with each other. When we measure the SST with instruments on ships and buoys, what we are mostly measuring is these changes, brought about by the oceans' complex and as yet poorly understood processes. As we go through the plots of temperature, it will become clear that the changes in temperature are greatly influenced by the waxing and waning of SSTs due, in particular, to the SOI and PDO index variations. These oscillations also affect land temperatures and the weather generally over much of the globe, including precipitation.

Volcanic activity

An external factor that affects temperature is volcanic activity (Figure 9.4, Table 9.1, page 156) (Ammann et al. 2003). Sulphur gases emitted by explosive volcanic eruptions into the stratosphere oxidise to produce sulphate aerosols. This process of gas-to-particle conversion has a time constant of around a month. Slowly the aerosol particles are deposited, with a time constant of around a year. Volcanic eruptions also discharge siliceous ash particles, but these, being larger (2 μm) than aerosols, leave the atmosphere within about three months. All of these products affect the radiation balance (Chapter 1), both in the solar and IR wavelengths. These events can affect temperature, but only for a short time, perhaps two or three years at most; for example, all temperature anomalies in 1991/2 were low (see graphs in Figures 9.6–9.12) due to the eruption of Mount Pinatubo (hereafter referred to simply as Pinatubo),

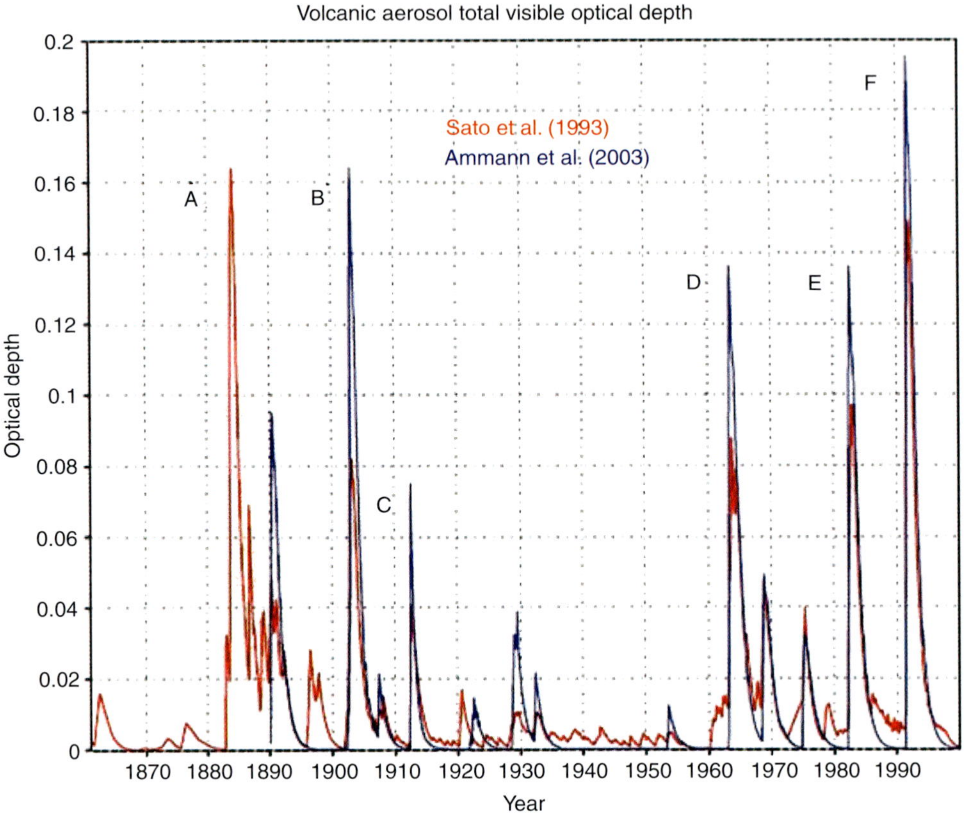

Figure 9.4 Graph of volcanic eruptions showing the optical depth estimates of sulphate aerosols in the stratosphere from 1860 to 2000 (reproduced from the IPCC AR4 2007). *Lettering* has been added to simplify identification, as follows: (A) Krakatau, Indonesia, 1883; (B) Santa Maria, Guatemala, 1902 and Pelée, Martinique, 1902; (C) Taal, Philippines, 1911; (D) Agung, Indonesia, 1963; (E) El Chicón, Mexico, 1982; (F) Pinatubo, Philippines, 1991. Information from Sato et al. 1993 (*red line*); Ammann et al. 2003 (*black line*). See also Table 9.1 (page 156), listing the known major volcanic eruptions since 79 AD.

which caused a fall of temperature of around 0.5°C and was of similar size to the Krakatau eruption of 1883. The years when major volcanic eruptions occurred are shown on the graphs of Figures 9.6a, b, 9.7a, b and 9.8c by highlighting the year of occurrence and the following two years.

Solar activity

While SST changes and volcanic activity have an undoubted and well-established direct effect on temperatures, the question of how, and to what extent, the Sun affects temperatures is 'work still in progress'. From a purely intuitive, and even logical, standpoint, the possibility of some involvement seems at least likely, for it is the Sun that warms the planet. Readers should refer to Chapter 1 for a more detailed analysis of the question.

Clouds

Clouds are a major area of uncertainty in climate research. They are poorly handled in climate models and are a primary cause of the differences between model 'projections' (predictions). The reduction of incoming solar energy by a small increase in low-level cloud cover would completely offset the increased energy retained by the enhanced greenhouse effect. As suggested in Chapter 1, although this is unproven, the amount of cloud might be affected by the number of galactic cosmic rays entering the atmosphere and thus by the Sun's activity. Clouds are also a product of evaporation and so are likely to increase if evaporation increases. An increase in cloud cover has been observed over the last 50 years, and this has also affected the *diurnal temperature range* (DTR) (the difference between the maximum and minimum temperature; dealt with in the section *Changes in maximum and minimum temperatures* at the end of this chapter). Clouds are thus a key unknown in climate research and it is difficult to quantify their effects. Some of the variability and trends seen in the following graphs may involve changes in cloud cover, but at present there is no way to be certain.

Changes in atmospheric circulation

Air temperature measured at any one point depends, in part, on where the air is coming from and its recent history. This is particularly so at mid to high latitudes, where depressions can alter the source of the air greatly in a day or hour as fronts pass through. As air circulation patterns change, so too does the source of the air. Both the source and the path the air takes affect its temperature, whether it comes across a warm sea or across snow. If atmospheric patterns change, it is likely there will be a shift of temperature. This could look like a change of climate if the change is persistent, such as a switch in the NAO from one state to another. Parker (2008) has developed a method to take account of this, not by using local wind direction as an indicator of air source, which was the method used earlier by Sexton et al. (2003), but by using a method of 'air-tracking' that examines sea-level barometric pressure fields in order to determine more precisely the source and path of the air and thereby deduce how much atmospheric circulation has affected the temperature anomalies. (See more on the use of this technique under the heading *Central England Temperature since 1659* below). In equatorial regions, air-tracking should be done using low-level wind direction, not pressure fields, because of the different nature of atmospheric circulation at the different latitudes.

Increased levels of carbon dioxide

The most familiar of all possible influences on long-term global temperatures is, of course, CO_2. Theoretically, an increase in CO_2 would, *if nothing else changed*, increase the amount of heat retained in the lower troposphere, and the atmosphere at the surface would warm (see Chapter 1 for the mechanisms and amounts). The physics of this is clear and simple. For greenhouse gases we must be thankful, for if it were not for their heat-retaining properties we would suffer a much colder climate than we do. Indeed life as we know it might not have emerged.

Whether or not an increase in temperature is an actual consequence of increased amounts of CO_2 is much less simple to ascertain, since in the real world many other factors are varying at the same time, such as amounts of cloud (and the unknowns that are unknown). Just as a fall in sunspot numbers, coincident with a fall in temperature at the Maunder minimum, does not necessarily confirm a connection between them, a rise in the amount of CO_2 coincident with a rise in temperature is also no proof of a connection. That is not to say it will not happen, but we have to recognise that it may not.

Changes introduced by changing site conditions and changing instruments

This matter has been dealt with at length in Chapters 2, 3, 4 and 5, and will not be repeated here. Suffice to say that, in addition to real changes in temperature, changes are also introduced by changes in the instruments used to measure the temperature and changes at the land sites where the measurements are made. Such changes introduce errors into the measurements which can be mistaken for real changes in the climate.

From the above, it will be clear that it is difficult to establish which of these many different processes are contributing to temperature changes at any one point in time, and their degree of contribution. With these various possible influences in mind, let us look at the actual changes in temperature that have occurred and try to establish some of the possible causes.

Hemispheric and global temperature changes since 1850

A selection of the CRUTEM3, HadSST2 and HadCRUT3 datasets is presented in this chapter in graphical form to illustrate how the temperature has changed over the past 155 years. As noted earlier, much more information is available in the form of anomalies for each 5° × 5° grid box.

In order to give an impression of the relative sizes of the changes in the two hemispheres and over the land and sea, the same scale has been used for all the graphs, even though this 'wastes space' where the changes are smaller; using different scales for each chart to fill all the space available could be misleading.

It has been generally agreed that there have been three major twentieth-century temperature trends: from about 1915 to 1945, from 1945 to 1975 and from 1975 to the present. As the charts show, while there are some differences in the timing, pattern and amplitude of the trends between the land and sea, and between the two hemispheres, they all have much in common. A comparison of the SOI and PDO index graphs (Figure 9.3a, b) with the temperature anomaly graphs also shows that these major temperature trends coincide with the major changes of phase and amplitude of the SOI and the PDO index (as marked by vertical lines in Figure 9.3a, b).

Table 9.2 summarises the trends (see page 170).

Land air temperatures

Land covers only 29% of the globe. Subtract the Antarctic and the parts of Africa and South America that are only patchily instrumented, and the area monitored is probably as little as one-fifth of the Earth's surface, mostly in the north. Land air temperatures also have greater variability from year to year than those of the sea (about six times as much in summer).

Northern Hemisphere land air temperatures

1975–2008 Let us start with the present and work backwards in time. Most of the warming since 1975 has occurred in the Northern Hemisphere (NH), over land, in winter and spring, at high latitudes (Figure 9.5) (Smith and Reynolds 2005). The warming is strongest over western North America, northern Europe and China in winter (DJF) and over Europe and northern and eastern Asia in spring (MAM). This is not reflected in the Southern Hemisphere or over the sea.

Figure 9.6a is a plot of CRUTEM3 annual land air temperature anomalies for the Northern Hemisphere, showing clearly the large temperature rise which has occurred since 1975, in part due to the anomalous high-latitude warming (Figure 9.5). The rise has a trend of +0.33°C per decade, the anomaly reaching a maximum of 0.8°C, individual years extending from −1°C to +1°C (expressed as anomalies from the 1961–90 mean).

It has been suggested that the urban heat island effect might be involved in some of this local anomalous seasonal warming (McKitrick and Michaels 2004) because it correlates with the location of many major cities and also occurs where satellite images show an abundance of artificial light at night

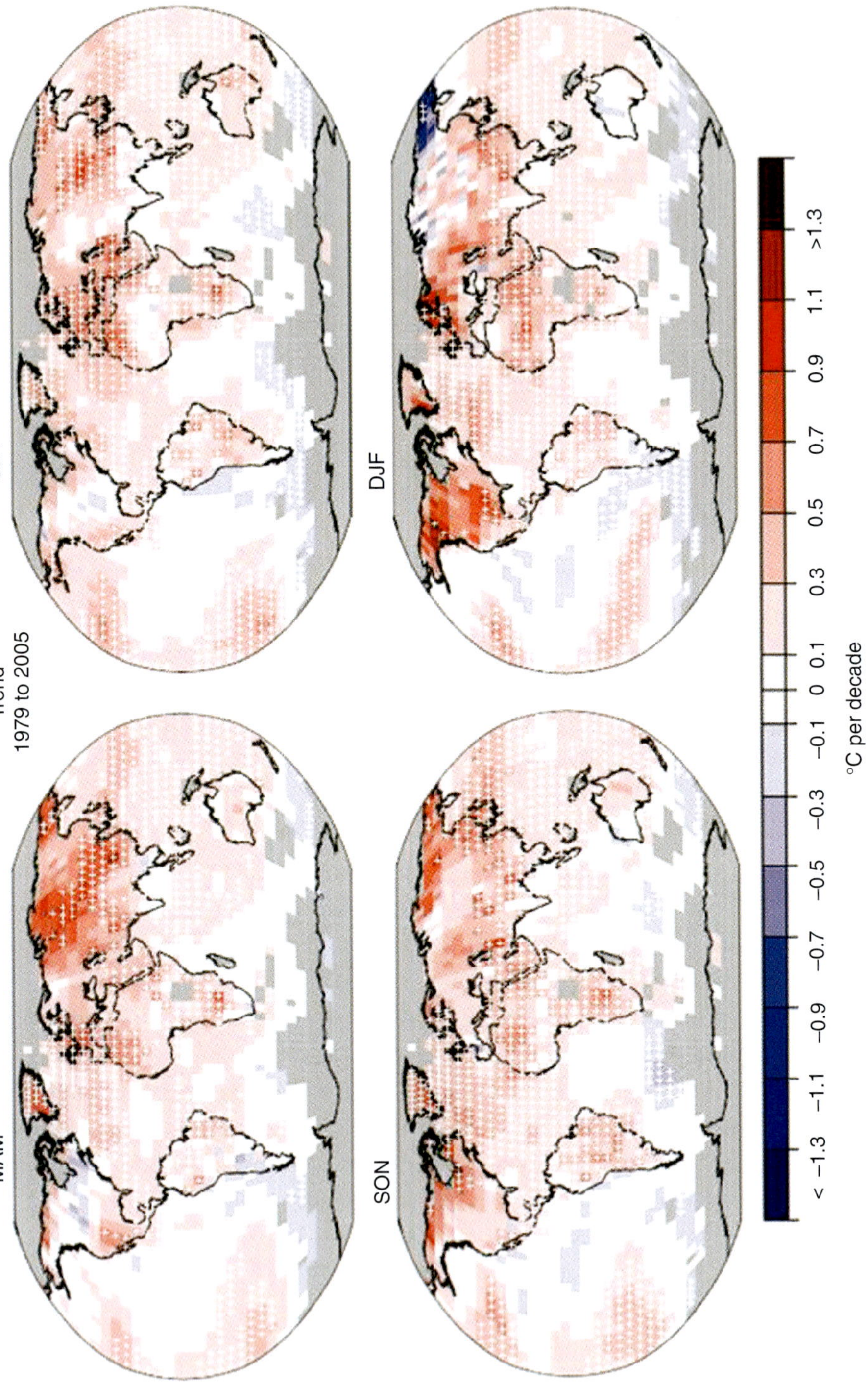

Figure 9.5 Maps showing the trend of temperature anomalies from 1979 to 2005 for the four seasons. See text for details. Areas with insufficient data are coloured grey (reproduced from the IPCC AR4 2007, updated from Smith and Reynolds 2005).

(see http://veimages.gsfc.nasa.gov//1438/earth_lights_lrg.jpg and the section *The urban heat island and land-use changes* in Chapter 4). But, again, association is no proof of causation and a change of atmospheric circulation has also been given as an explanation. De Laat and Maurellis (2006) suggest that changes in land use or in cloud cover around areas of industrialisation could be responsible. They see the changes as anthropogenic, but not due to CO_2. Climate models are also said to predict that warming will occur in these regions due to increases in CO_2. At the end of this chapter we also look at the possibility of changes in the diurnal temperature range (the difference between maximum and minimum daily temperatures) playing a part in the warming, and in Chapter 10 we will examine what is happening in the troposphere in these high northern latitudes in winter and spring.

However, much of the increase in temperatures post-1975, and its timing, is due to a switch from a series of large and frequent La Niña (cold) events, with fewer and smaller El Niño (warm) episodes, to the reverse after 1975 (Figure 9.3a). The start of this climate shift is marked with a solid vertical line. At around the same time (1975) the PDO also switched to a warm cycle, following its three decades in a negative (cold) phase (Figure 9.3b). It is probable that we need look no further than these for an explanation of most of the trends post-1975, not only in the NH land but globally, with perhaps a few embellishments arising from volcanic activity. Changes in cloud cover might also be involved. These oscillations account for the general timing of the warming, but not for the extra warming in the high northern latitudes in winter; we look at this again later.

Regarding volcanic activity and its effect on temperatures, the slightly lower anomalies in the early 1980s (Figure 9.6a *E*) might be due to the eruption of El Chichón, while the lower anomalies of the early 1990s (Figure 9.6a *F*) were certainly due to Pinatubo (see Table 9.1 and Figure 9.4).

1945–1975 Prior to the recent warming there was a fall in temperature between 1945 and 1975 and again we need only to look to the SOI and PDO index variations for the most likely cause. From Figure 9.3a it can be seen that after a very large El Niño in 1940/1 (left of the dashed line), when the temperature anomaly rose to, and above, the 1961–90 mean, there was then a switch to frequent and large La Niña events, with few El Niño phases, right up to the 1975 switch to warmer times. In similar fashion (Figure 9.3b, dashed line) the PDO reversed suddenly from a warm to a cold phase in 1943, where it remained up to 1975.

In 1963 the eruption of Agung in Indonesia probably caused the lower anomaly around that time (Figure 9.6a *D*).

It is interesting to look, with hindsight, at how this fall in temperature gave rise in the 1970s to some concern about a possible new Little Ice Age starting. Even in 1982 it was still not clear to climatologists whether the temperature was still falling

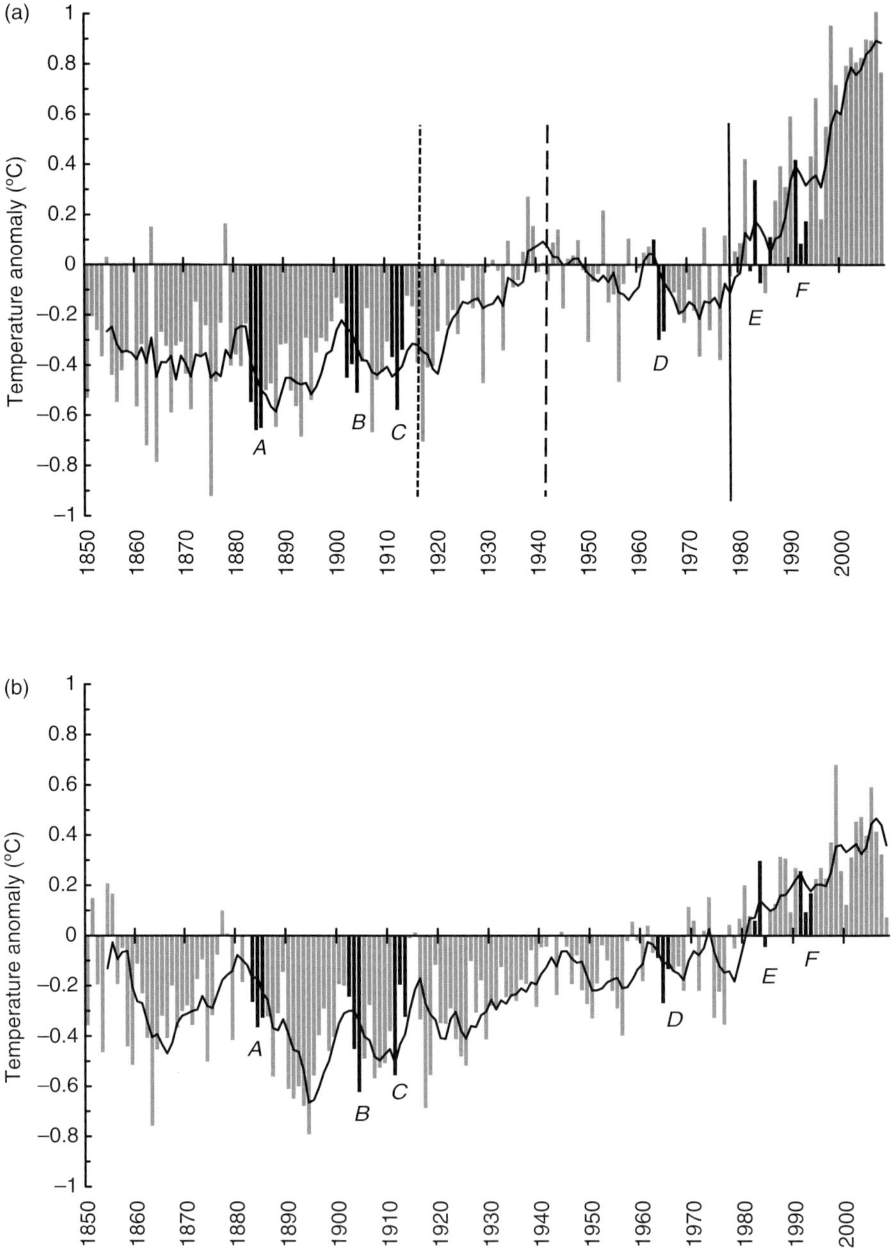

Figure 9.6 (a) Land air temperature, Northern Hemisphere; (b) Land air temperature, Southern Hemisphere; (c) Land air temperature, global. In a and b the six major volcanic eruptions are marked as darkened columns *A–F*, as in Figure 9.4. See text for discussion.

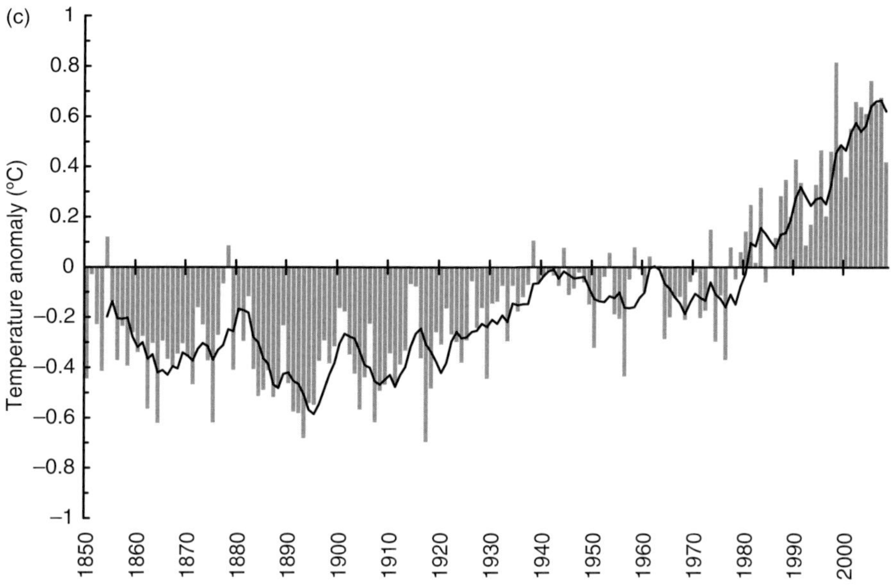

Figure 9.6 (*cont.*)

(post-1945), or whether it had levelled off, or was perhaps just starting to rise (Kelly et al. 1982). In the early 1980s only temperatures in the NH had been analysed, and it was only what is shown in Figure 9.6a (and only up to 1980) that was the picture currently available, at a time when the temperature had been falling for 30 years. Someone in 1980 looking at such a graph might reasonably have feared a continued fall. However, the downward trend puzzled climatologists, as even in the 1970s professional researchers were expecting a rise in temperature due to increasing CO_2 levels in the atmosphere (which C.D. Keeling had been measuring since 1958 at the Mauna Loa observatory in Hawaii (Keeling and Whorf 2005)). The fact that the opposite was happening is a reminder that many factors are in operation in driving the climate system.

However, Peterson et al. (2008) have pointed out that most of the papers written at that time were concerned with the possible future warming due to CO_2 increases and that far fewer papers were concerned with the cooling. It was the media that picked up the cooling because the media are mostly concerned with what is happening right now.

An intriguing aside about this episode is that the reasons given for the expected warming and the observed cooling were both entirely anthropogenic (Peterson et al. 2008). The observed cooling was explained as possibly being due to increased aerosols from industry, power generation and cars, reflecting more sunlight back to space, while the anticipated warming was due to increased carbon dioxide from

Table 9.1. *Major volcanic eruptions*

Name	Country	Year
Vesuvius	Italy	79
Etna	Italy	1169
Santorini	Greece	1470
Kelut	Indonesia	1586
Vesuvius	Italy	1631
Etna	Italy	1669
Ringgit	Indonesia	1686
Papandayan	Indonesia	1772
Laki	Iceland	1783
Unzen	Japan	1792
Tambora	Indonesia	1815
Krakatau	Indonesia	1883
Santa Maria	Guatemala	1902
Pelée	Martinique	1902
Taal	Philippines	1911
Kelut	Indonesia	1919
Merapi	Indonesia	1930
Rabaul Caldera	Papua New Guinea	1937
Lamington	Papua New Guinea	1951
Hibok	Philippines	1951
Agung	Indonesia	1963
St Helens	USA	1980
El Chichón	Mexico	1982
Nevado del Ruiz	Columbia	1985
Lake Nyops	Cameroon	1986
Pinatubo	Philippines	1991
Unzen	Japan	1991
Mayon	Philippines	1993

Note: The effect of an eruption on temperature is short-lived, lasting for about two years. The major volcanic eruptions are listed here with dates to allow comparison with the graphs of temperature anomalies. This can be compared with the graph of eruptions in Figure 9.4.

industry, power generation and cars, trapping more IR radiation. We seem, as a species, always to see ourselves as being at the centre of things.

1910–1945 There was an earlier rise of temperature starting around 1910, peaking in 1940. The general view is that anthropogenic influences at this time were negligible, so we can assume that most were natural. This early trend is also likely to have been dominated by the two main teleconnections, as a glance at Figure 9.3a, b confirms. The SOI shows that around 1915 (dotted line in Figure 9.3a) a large El Niño raised the temperature toward the mean, thereafter there being a sudden transition to smaller El Niño and larger La Niña episodes, causing the temperature to drop suddenly. Over the next 25 years the magnitude of the La Niña events slowly reduced, such that by the mid-1930s the temperature anomaly was back to the mean, being followed by the large 1939–42 El Niño (left of the dashed line in Figure 9.3a) that took the temperature anomaly to 0.2°C above the 1961–90 mean in 1938, after which the slow fall occurred, as already discussed above. A similar pattern is seen in Figure 9.3b, a slowly increasing positive trend in the PDO index (to the right of the dotted line) adding to the effect of the rise in the SOI.

Pre-1910 Prior to the rise from 1915, the anomalies fluctuated considerably, with nearly equal-sized El Niño and La Niña episodes alternating. During this period the temperature anomalies oscillated between the mean (or near to the mean) and fairly low individual values.

Volcanoes were quite active in this period, with three major eruptions (Figure 9.6a *A*, *B*, *C*). These each clearly lowered the temperature for a couple of years following the events, but from this graph it would not seem that volcanic activity has a very great influence on temperatures. However, should there be a series of major eruptions one after the other, the effect would become more noticeable.

Over the full period (1850–2008) the trend was 0.063°C per decade.

Southern Hemisphere land air temperatures

1975–2008 Figure 9.6b is a plot of the CRUTEM3 annual land air temperature anomalies for the Southern Hemisphere (SH), where changes are roughly similar to changes north of the equator, but with quite notable differences, the most prominent being the much smaller rise post-1975 (0.17°C per decade; half that in the north, of 0.33°C per decade) probably because there was no extra warming comparable to that in the Northern Hemisphere's high latitudes in winter (for whatever reason this occurred). One can speculate that it was because of the lack of large population centres and industry in the SH countries, but that would be pure speculation, although not an unreasonable hypothesis. Also there is much less land.

1895–2008 Another noticeable difference in the SH is the nature of the temperature increase which, although generally accepted as having starting around 1915–20 (in the NH) is clearly part of a longer, intermittent rise that could be seen as having started as early as 1895, with periodic returns to the 30-year mean (at 1915, 1945, 1960, 1970 and 1980), with large negative phases in between but of slowly reducing magnitude. The overall effect is that of a continuous rise over the 110 years from around 1895. There is not the noticeable fall between 1945 and 1975 clearly evident in the NH, just a slowing of the increase. The trend since 1975 thus appears more as a continuation of that started in 1895. This effect is enhanced by the post-1975 rise being only about half that in the NH. Over this century-long period, there was a trend of 0.09°C per decade. Over the full period (1850–2008) the trend was 0.036°C per decade, half that in the NH.

However, reference back to Figure 9.6a suggests the rise in the NH can also be seen as starting as early as the mid-1880s.

Volcanic activity is lettered in the same way as in Figure 9.4.

Global land air temperatures

Figure 9.6c is a plot of the CRUTEM3 annual land air temperature anomalies for the globe. Being a merging of the two hemispheres (in the weighted proportions 68% NH, 32% SH, to allow for the different land areas in the two hemispheres) the end-product has characteristics of both, but biased towards the NH. The recent rise in temperature (post-1975) in the combined dataset has a trend of 0.24°C per decade, while over the whole period it was 0.048°C per decade.

Sea surface temperatures

Temperature trends over the sea are not too dissimilar to those on land. Rather we should say, perhaps, that land air temperatures show similar trends to those over the seas, modified by the more ephemeral, transitory processes in our comparatively thin atmosphere. Because of the large areal extent and great thermal inertia of the oceans, land air temperatures will tend to follow the sea surface temperature, not vice versa. Close to the ground, where we make measurements, the local temperature will obviously be modified by surrounding conditions, especially in the middle of large continents less so nearer the coasts.

Northern Hemisphere sea surface temperatures

Figure 9.7a is a plot of HadSST2 annual sea surface temperature anomalies for the NH, the three twentieth-century phases of temperature change being similar to those over the NH land but smaller and with different timings.

The first stage of warming clearly started around 1910–15 (not earlier as over land), from the lowest anomaly of −0.6°C, in a more definite way than on land. The rise was rapid and large (0.19°C per decade), peaking in the early 1940s. This rise, being largely free from anthropogenic influence, demonstrates that an increase of this size can occur naturally. The fall from 1945 to 1975 is less obvious than that over land, being a series of temporary dips below the mean. Post-1975 warming is about half that over the NH land (0.20°C compared with 0.33°C per decade) and about the same as the rise from 1915 to 1975. These trends can be explained through the same changes in the SOI and the PDO index as discussed with regard to the NH land. The overall trend since 1850 is about half that over the land (0.033°C compared with 0.063°C per decade). There has been a fall in temperature over the last four years because the El Niño phases are declining, while La Niña increases.

Volcanic activity is lettered as in the previous land figures (Figure 9.6a, b).

Southern Hemisphere sea surface temperatures

Figure 9.7b is a plot of the HadSST2 annual sea surface temperature anomalies for the SH. It shows some similarity to the northern ocean trends up until the mid twentieth century, but the fall around 1943 is steep, sudden and large (occurring in just one year: 1946). This may be explained by the sudden switch from the large El Niño in the early 1940s to predominantly large La Niña events, gradually tempered and offset by slowly increasing El Niño events. I am informed, however, that this sudden drop in 1945 is considered by some researchers to be due to an error in the data, brought about by a change in the way in which British and American ships measured SST during and immediately after the Second World War, although nothing has yet been published on the matter.[1] If correct, the effect would be to delay the fall until about the mid-1950s, thereafter rejoining the present curve. If this is the case, it demonstrates just how vulnerable trends can be to how the measurements are made and how important it is to improve and standardise how measurements are made in the future. But there is doubt here, because the drop was much greater in the SH than in the NH seas while, in addition, there was also a similar sudden fall over the SH land which will not have been due to an instrument error and which is not present in the NH land record. So instrument error may not be the answer; the fall could be real and this is supported by a look at the SOI and PDO index graphs (Figure 9.3 a, b). If real, the overall effect is that of two trends. The first was a large rise from 1910 to 1940 of 0.6°C (0.2°C per decade) (this is not in question), followed by a sudden drop of 0.4°C in a year, in turn followed by a second rise of 0.7°C from 1945 to 2000 (0.13°C per decade). There has been a noticeable drop in temperature since 2002 (see later).

It will be clear from the foregoing that attributing causes to the ups and downs of each anomaly, or run of anomalies, is a matter requiring close scrutiny; even then there is much uncertainty and guesswork.

Volcanic activity is again shown darker, marked *A–F*. It does not have as much influence on SST as it does on land air temperatures because SST changes slowly and cannot respond as much to transitory influences.

Global sea surface temperatures

Figure 9.7c is a plot of the HadSST2 sea surface temperature anomalies for the globe. As this is a merging of the two hemispheres (weighted $0.44 \times NH + 0.56 \times SH$) the

[1] But see Thompson et al. (2008).

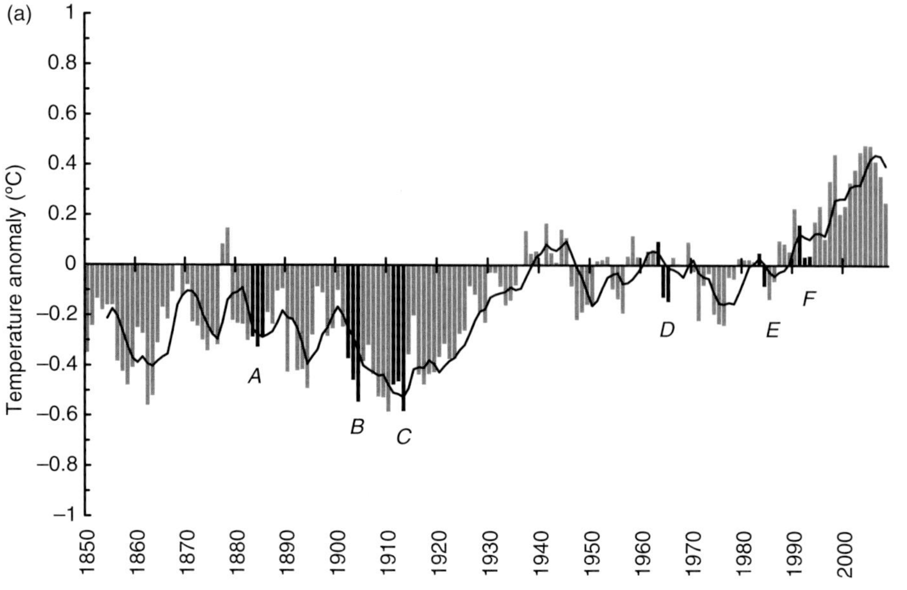

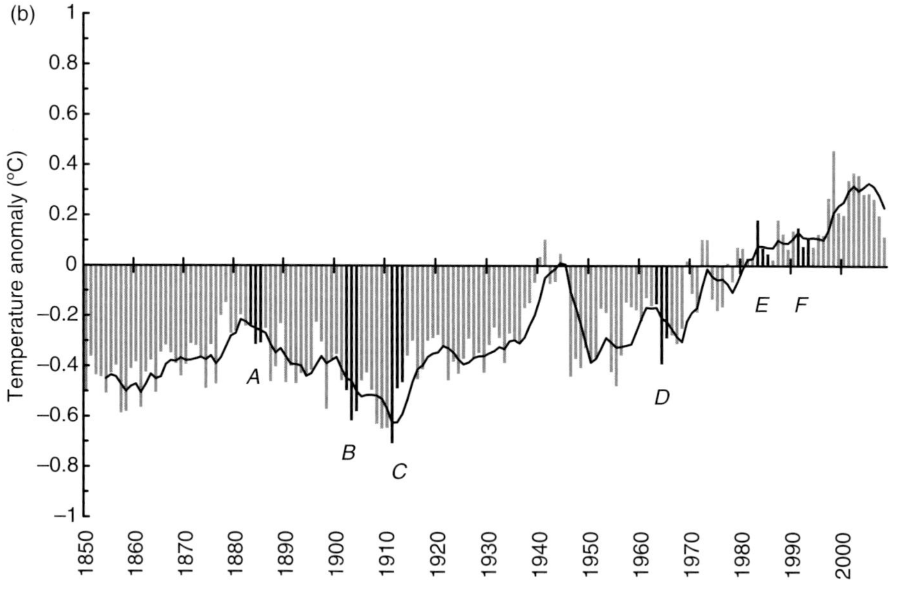

Figure 9.7 (a) Sea surface temperature, Northern Hemisphere; (b) Sea surface temperature, Southern Hemisphere; (c) Sea surface temperature, global. See text for details. In a and b the six major volcanic eruptions are marked as darkened columns *A–F*, as in Figure 9.4. See text for discussion.

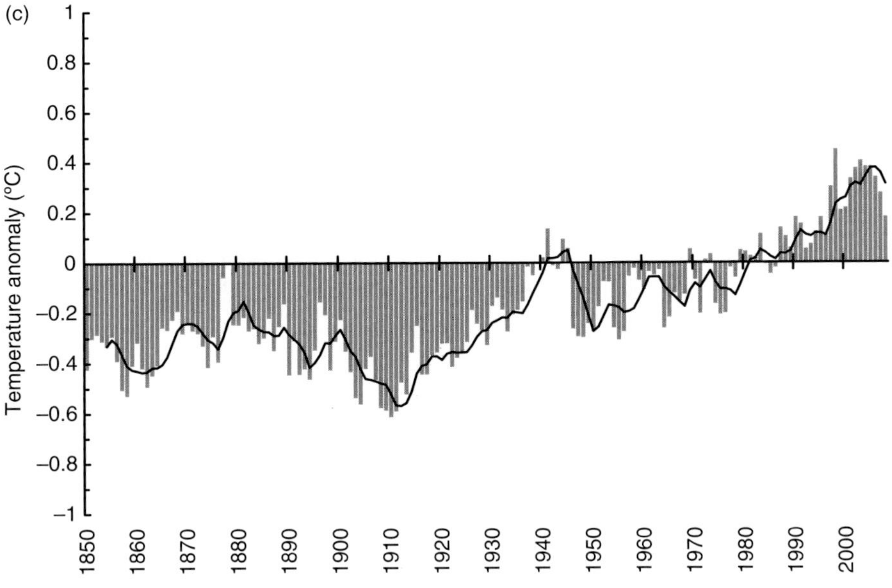

Figure 9.7 (*cont.*)

end-product is very much like the SH plot with the post-1975 rise being 0.16°C per decade compared with 0.24°C per decade over the global land. Over the full 155 years, the upward trend was 0.038°C per decade. The post-1940 fall is tempered by the smaller fall in the NH.

Land-air and sea-surface temperatures combined

The following three graphs (Figure 9.8a, b, c) are included for completeness only. They hide too many details to be of use for anything other than a simple answer, and such merging is also questionable as already explained in Chapter 8.

Northern Hemisphere land-air and sea-surface temperatures

Figure 9.8a is a plot of HadSST2 for the NH land-and-sea anomalies combined. The trend since 1975 was +0.24°C per decade, the overall trend from 1850 was 0.044°C per decade. The other two 'classical' trends, 1915–45 and 1945–75, are clearly seen, preceded by a slow undulating fall from 1850 of about 0.25°C, with much variability, including a brief warm spell in the late 1870s.

Southern Hemisphere land-air and sea-surface temperatures

Figure 9.8b is HadSST2 for the SH land-and-sea temperature anomalies combined. The sudden drop in the SH SST around 1945 has been echoed in this combination of land and sea in the south because there was a similar but smaller fall over the

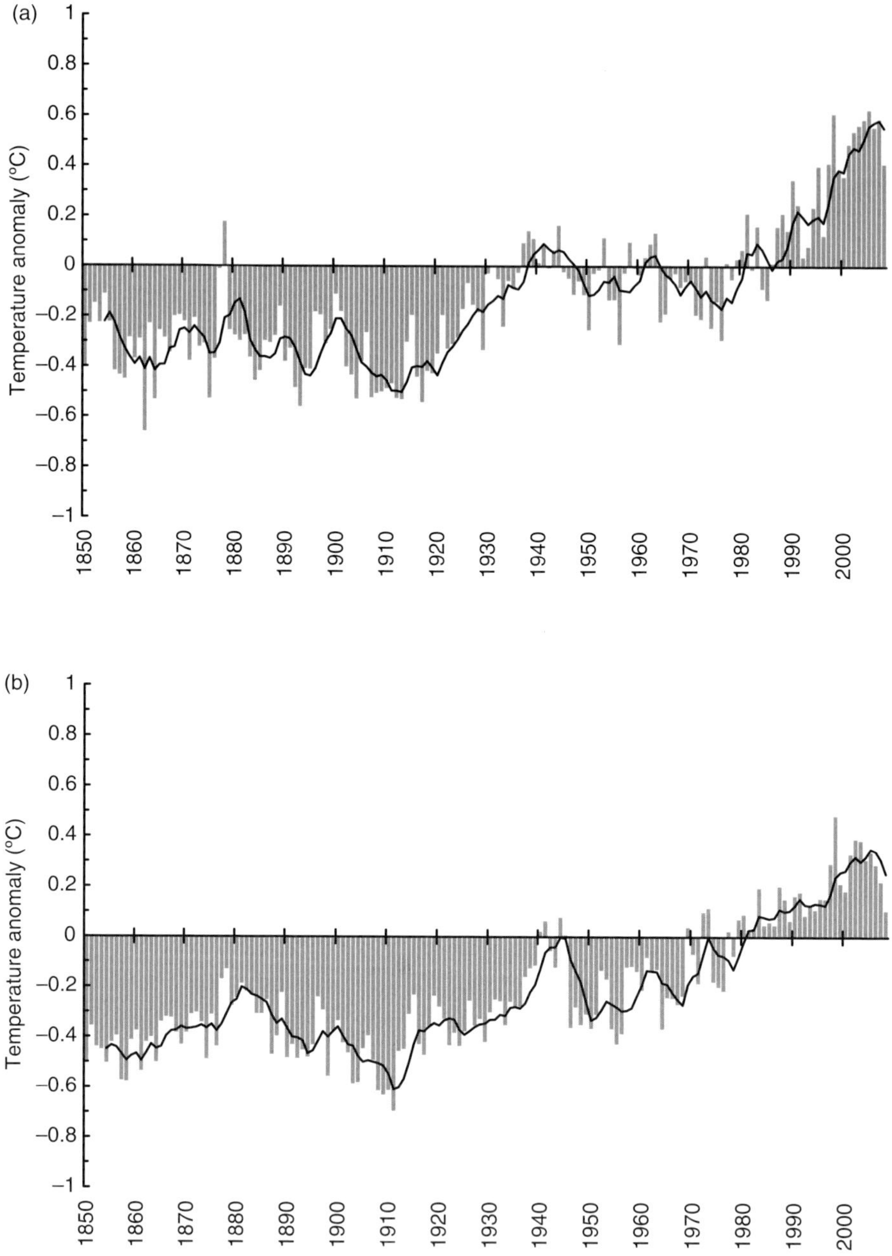

Figure 9.8 (a) Land air and sea surface temperatures combined, Northern Hemisphere; (b) Land air and sea surface temperatures combined, Southern Hemisphere; (c) Land air and sea surface temperatures combined, global. In c, the six major volcanic eruptions are marked as darkened columns *A*–*F*. See text for discussion.

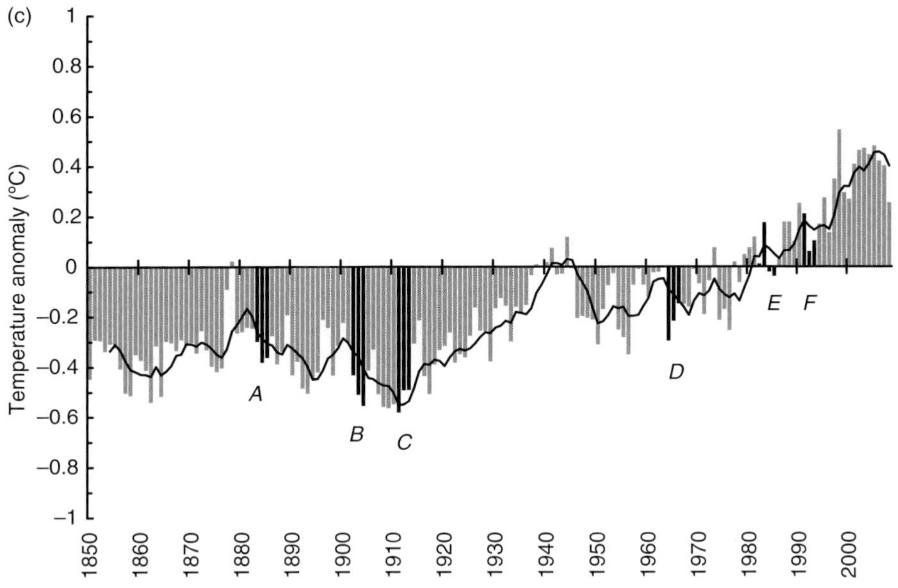

Figure 9.8 (*cont.*)

SH land. The post-1955 rise of the trendline to the present day is almost linear, the linear trend after 1975 being 0.11°C per decade. Over the full period, the trend was 0.042°C per decade.

Global land-air and sea-surface temperatures

Figure 9.8c is HadSST2 land-and-sea, NH and SH temperature anomalies combined. This integrates all the anomalies from the two hemispheres, for the land and sea, combined into one figure. The post-1975 upward trend was 0.19°C per decade, with a slight fall over the last few years. The overall trend from 1850 was 0.041°C per decade. The volcanic events are marked. They do not have a great impact.

Recent temperature changes

The 150-year-long graphs in Figures 9.6, 9.7 and 9.8 telescope all the short-term detail. Since it is the recent changes that are of greatest concern, because humanity might be responsible for some of them, it is useful to look more closely at the changes since 1975 by re-plotting the most recent 35 years of the CRU/Had datasets.

Land air temperatures

Northern Hemisphere land air temperatures Figure 9.9a shows the temperature changes (anomalies relative to the 1961–90 mean) during the last 35 years over land

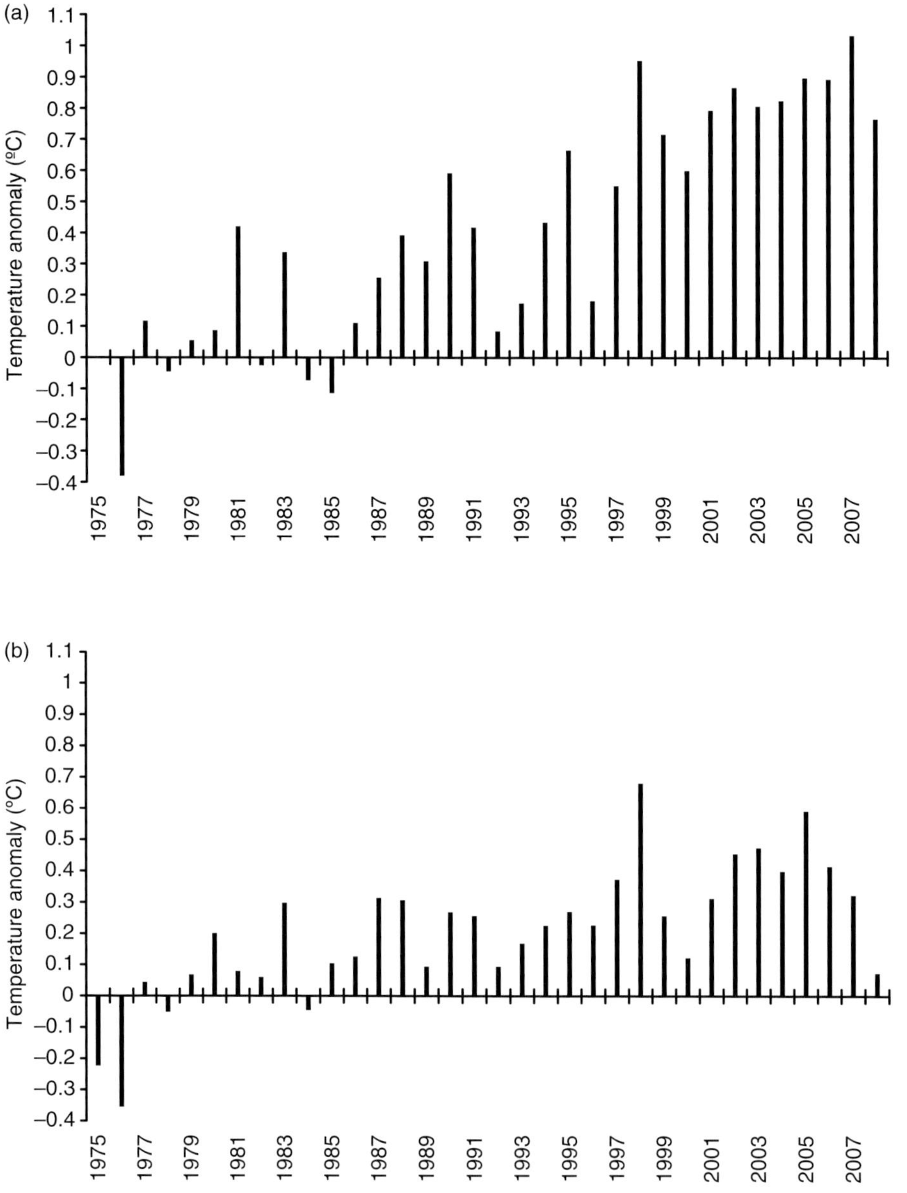

Figure 9.9 (a) Land air temperature, Northern Hemisphere, since 1975; (b) Land air temperature, Southern Hemisphere, since 1975; (c) Land air temperature, global, since 1975. See text for discussion.

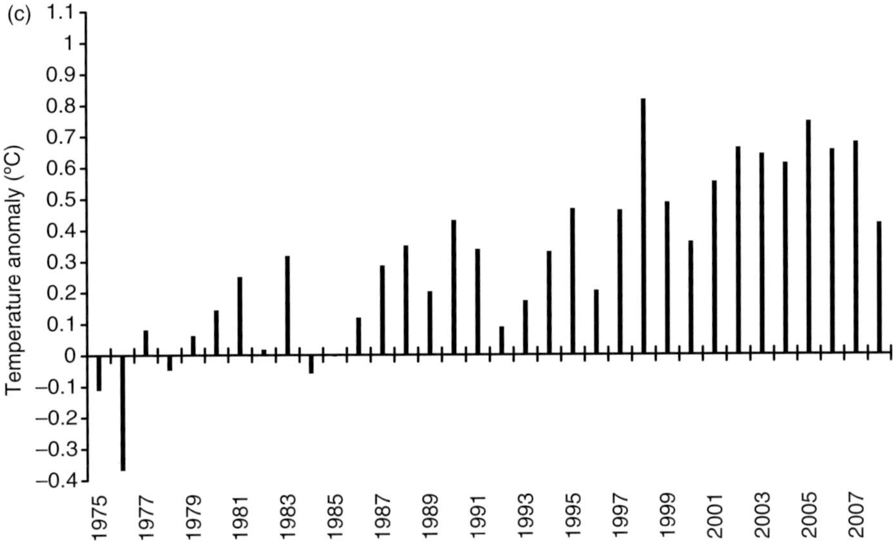

Figure 9.9 (*cont.*)

in the NH, where most warming has occurred (during the winter at high latitudes, as explained earlier; see Figure 9.5). The graph shows that, with the exception of 1981 and 1983, the temperature remained at or near the 1961–90 mean until around 1986, a clear upward trend then starting, albeit with some variability, 1992/3 being low due to the eruption of Pinatubo, and 1998 being the second warmest. This especially warm year is seen in all of the other graphs and is due to a particularly large El Niño event. Unique to the NH land is the even warmer year of 2007, which is not explained by an El Niño. This is not, however, reflected in the SH or in the SST records (see later). Apart from the 2007 peak, the temperature has been fairly steady since 2002, with no major trend occurring over the last six years.

Southern Hemisphere land air temperatures Temperature change south of the equator (Figure 9.9b) is notably less than in the north. The large rise in the NH from 1986 is not echoed in the south. It is difficult to pinpoint exactly when a rise started, since frequent low anomalies appear regularly. In contrast to the NH, 2007 was not the warmest in the south, being cooler than 1998 and similar to many earlier years, 2005 being the second warmest year in this half of the world, probably due to the largest El Niño since 1998. If the 1998 event is omitted, the trend in the SH land temperature is moderate and gradual, starting in the late 1970s. Apart from 2005, however, the SH lands have cooled since 2003.

Global land air temperatures The amalgam of the NH and SH anomalies (weighted in the proportions 68% NH, 32% SH) produces Figure 9.9c. From about 1976 there was a gradual upward trend in temperature to the present but with several cooler years interspersed. This combined chart shows a more or less constant temperature after 2002, with 2008 being lower.

Sea surface temperatures

Northern Hemisphere sea surface temperatures Figure 9.10a reveals the very different temperature pattern from that in the NH land. Not only is the amplitude of the changes much smaller, but until about 1985 it could be said that there was no obvious trend, with temperature anomalies remaining around the 30-year mean. Indeed, with the exception of 1990 and 1991, the main trend could be seen as starting in 1994, peaking in 2004, followed by a downward trend over the last four years. The 1998 large anomaly is also evident as usual. The 1998 event demonstrates clearly the imprint of the El Niño and La Niña events on the temperatures.

Southern Hemisphere sea surface temperatures The oceans of the south (Figure 9.10b) show a small, fairly constant anomaly of under ±0.2°C relative to the 30-year mean from 1975 to 1996, with no apparent trend. There then followed a modest increase up to 2002 (with the El Niño year of 1998 again standing out as being unusual). From 2002 onwards the trend was downwards. The seas of the Southern Hemisphere, therefore, show little warming and have been cooling for the last half decade.

Global sea surface temperatures Combining SST measurements, north and south (weighted $0.44 \times NH + 0.56 \times SH$), produces Figure 9.10c. Since the temperatures of the seas, both north and south, follow somewhat similar trends, their global behaviour reflects this, showing no upward trend (from the 1961–90 mean) until around 1987 or even 1992. From around 1992 there was a slow consistent rise, peaking in the ubiquitous 1998 event, and falling since 2003. Since the oceans cover 71% of the globe and behave similarly north and south, it can be argued that this figure is the closest representation of the Earth's recent temperature change that we have, uncontaminated by large inter-annual differences of uncertain origin in the land data. But I am obliged to finish by adding in the land measurements.

Land and sea temperatures combined

Northern Hemisphere land and sea temperatures combined Figure 9.11a shows no trend in the NH combined land and sea anomalies until around 1987, the anomalies remaining close to the 1961–90 mean until then. From then on, with the exceptions of 1992 and 1993 (low in all the charts, land and sea alike, due to the eruption of

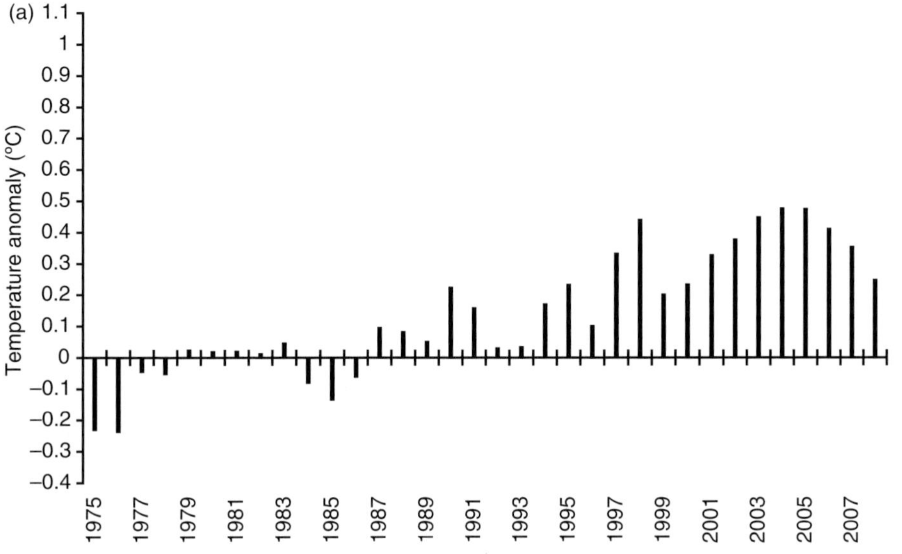

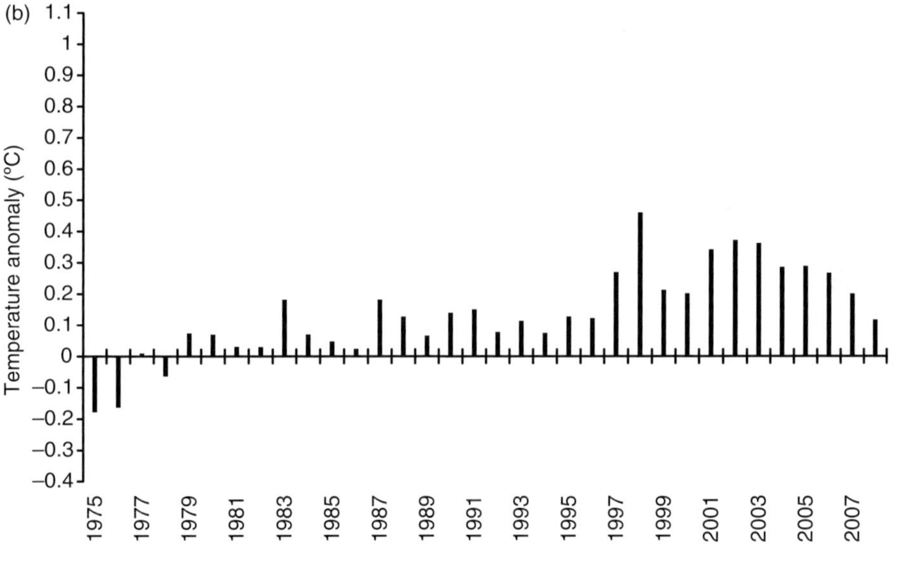

Figure 9.10 (a) Sea surface temperature, Northern Hemisphere, since 1975; (b) Sea surface temperature, Southern Hemisphere, since 1975; (c) Sea surface temperature, global, since 1975. See text for discussion.

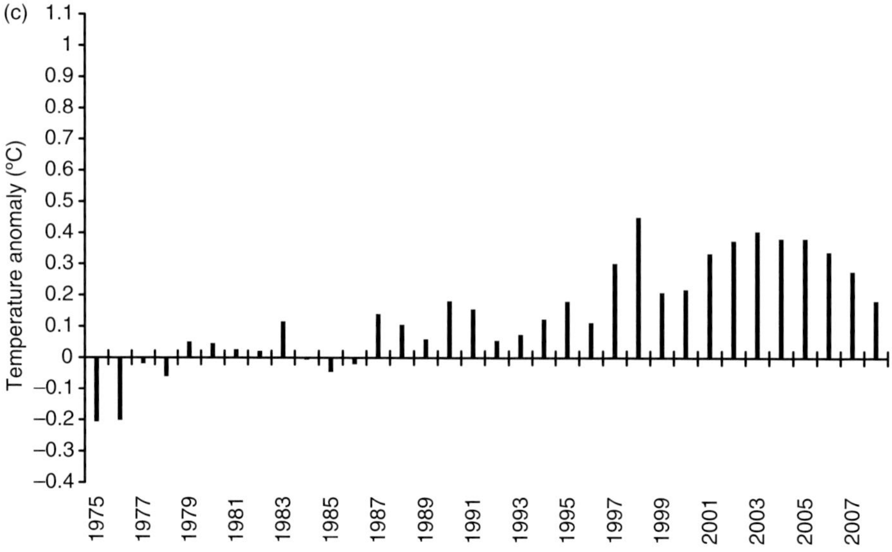

Figure 9.10 (*cont.*)

Pinatubo), there is an upward trend peaking in 2005, just marginally higher than the all-pervasive 1998 peak, since when the temperature has fallen slightly.

Southern Hemisphere land and sea temperatures combined Figure 9.11b shows the much more muted behaviour in the Southern Hemisphere of land and sea combined. While there is a gradual upward trend from (say) 1975, it is very small and hardly worthy of being called a 'trend'. If things remained as they were until 1997 we could not say there was much, if any, warming worth speaking of. Even then the warming was gentle and short-lived, peaking at around +0.4°C in 2002 with a gentle fall thereafter. The Southern Hemisphere, land and sea combined, has not warmed much and is now back near the 1961–90 mean.

Global land and sea temperatures combined As Figure 9.11c shows, the globe as a whole, land and sea combined, could be said to have warmed from around 1975, with intermittent cooling until, perhaps, as late as 1987, a peak being reached as usual in 1998. This maximum was not subsequently exceeded and since around 2002 the temperature has been steady, falling slightly since 2005. Doubtful as I am about the efficacy of merging everything into one figure, if we accept the procedure, Figure 9.11c presents a global picture of events over the last three decades.

While the present stable and cooling pattern may well not continue, we are not experiencing a runaway exponential rise either.

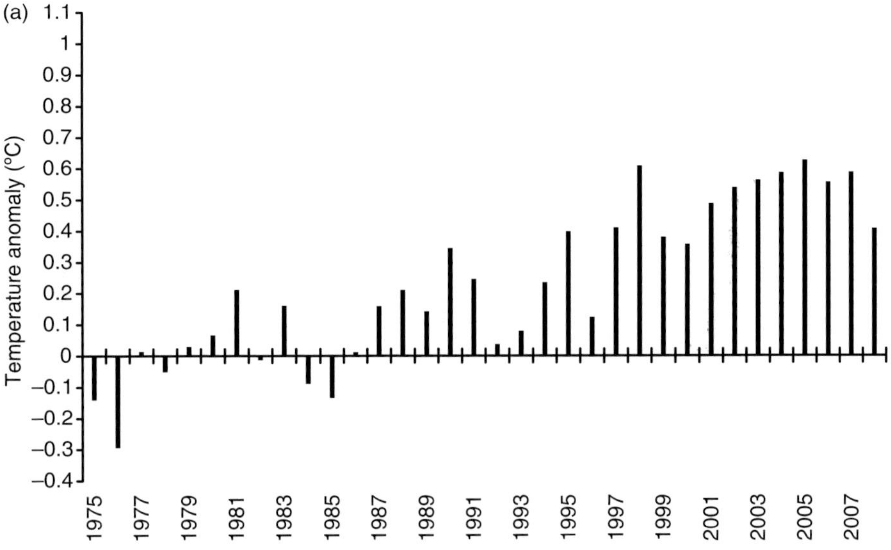

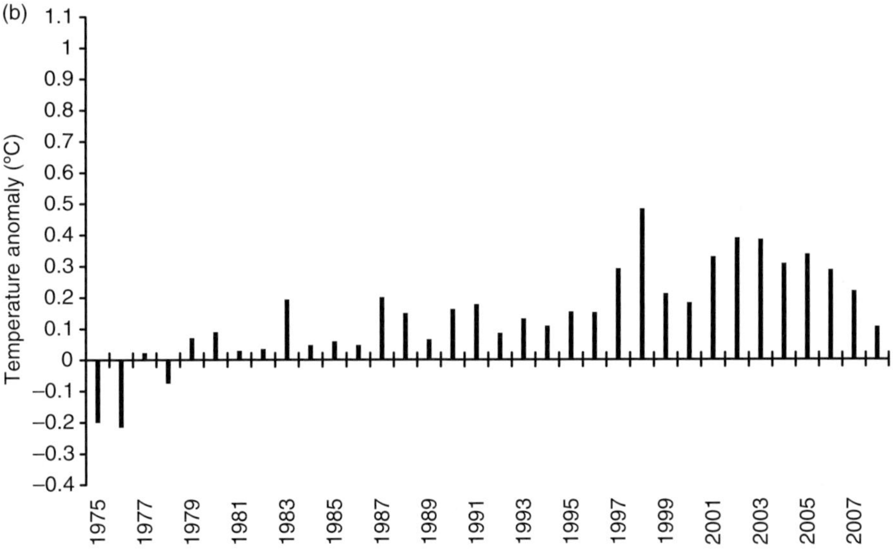

Figure 9.11 (a) Land air and sea surface temperature, Northern Hemisphere, since 1975; (b) Land air and sea surface temperature, Southern Hemisphere, since 1975; (c) Land air and sea surface temperature, global, since 1975. See text for discussion.

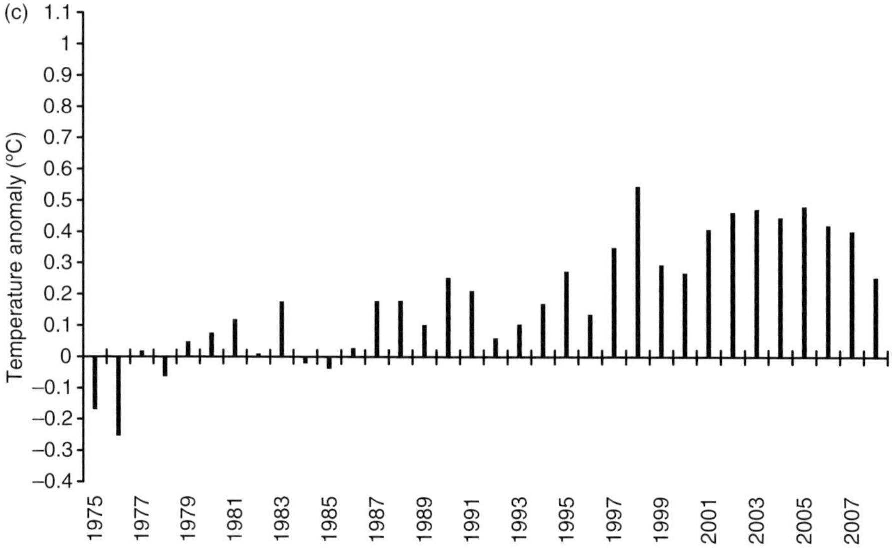

Figure 9.11 (*cont.*)

Table 9.2. *Temperature trends (approximate only)*

	Post-1975 trends (°C per decade)	1850–2005 trends (°C per decade)
LAND		
Northern Hemisphere	0.33	0.063
Southern Hemisphere	0.17	0.036
Globe	0.24	0.048
SEA		
Northern Hemisphere	0.20	0.033
Southern Hemisphere	0.12	0.042
Globe	0.16	0.038
LAND AND SEA COMBINED		
Northern Hemisphere	0.24	0.044
Southern Hemisphere	0.11	0.042
Globe	0.19	0.041

Note: The temperature trends for the recent (post-1975) changes and for the trends over the whole instrumental period (1850 to the present) are given here to aid in comparing the difference in trends over the land and sea and the two hemispheres. Trends are derived from the individual anomalies not from the trendlines. They are approximate only.

Central England Temperature since 1659

Because there were few reliable temperature measurements prior to 1850, we have no precise quantitative knowledge of earlier global temperatures, although inferences have been made by using various forms of proxy measurements such as tree rings, ice and ocean-sediment cores and historical documents, culminating in the 'hockey-stick graph' (Mann 2001). These will not be discussed here because they are not reliable global figures. The matter involves discussion about the Little Ice Age and the Medieval Warm Period, whether they were global or regional events, and what might have caused them (Jones 2008). This is a complex matter that will not be investigated here (but see page 15).

The dataset

In England, however, regular and careful instrument measurements of temperature began to be made as soon as thermometers had been invented in the mid seventeenth century (see Chapter 2) and the dataset known as *Central England Temperature* (CET) was compiled from these early measurements. The CET record was originally published by Gordon Manley (1953, 1974) and subsequently extended and updated by Parker et al. (1992), Parker and Horton (2005). It is representative, roughly, of a triangular area bounded by Bristol, Lancashire and London. This record represents the longest series of monthly temperature observations in existence. It is monthly from 1659, and a daily version has been produced from 1772. The earliest years, 1659–1722, only have monthly means given to the nearest degree or half a degree (Fahrenheit). The record is currently compiled from stations at Rothamsted, Malvern and Stonyhurst, although there have been several changes during its long history (see Parker et al. 1992, 2005 for details of the site changes).

While the CET record only tells us about events in central England, it throws useful light, nevertheless, on what happened here prior to 1850, during the last 360 years. Being only regional gives CET some advantage because the dataset does not involve the merging of readings from very different climates and hemispheres; a procedure, as readers will have gathered by now, I (rightly or wrongly) have doubts about.

I will not address the question of how 'accurate' the CET series is; sources of error will be the same as those already discussed in Chapters 2, 3 and 4. Parker et al. (1992, 2005) and Parker (2008) describe in detail the efforts made to correct the readings as much as is practicable. Let us assume that the measurements are as good as can be obtained and see what they tell us about the temperature over England during the last 350 years or so. The datasets can be downloaded from http://hadobs.metoffice.com/hadcet/data/download.html.

CETs are presented as absolute temperatures, not anomalies referred to any 30-year period (all the other data considered in this chapter are compared with a

30-year mean). To make CETs compatible with these other graphs, I have converted them to anomalies referred to the same period as the CRU/Had datasets; that is to 1961–90. The results are shown in Figure 9.12a, b, c.

Temperatures from 1659 to 2008

Figure 9.12a shows the full run of the CET dataset from 1659, expressed as anomalies from the 1961–90 mean. A feature that is immediately obvious is the very low temperatures at the start of the record. This period coincided with the *Maunder minimum* (MM), a part of the Little Ice Age in northern Europe, when sunspots were almost non-existent (see Chapter 1). The MM started around 1650 and began to come to a close around 1700, ending by 1715. But association is no proof of causation, and we cannot say that this period was cold because of low solar activity. At the same time, however, neither can we exclude the possibility that it was; we need a better understanding of solar–climate interactions. Recent research has shown, for example, that the Sun has become increasingly active over the last few hundred years and that over the last 60 years it has been more active than any time over the last 1,159 years. It is quite possible, of course, that several factors were influential; perhaps volcanic activity also played a part.

Between 1700 and 1710 there followed a very rapid 1.5°C rise, followed by a slight fall to 1715 and then a further rise of 0.5°C, peaking around 1735, when temperatures rose to above the 1961–90 mean for a decade or more, comparable to temperatures in the 1990s. The temperature increase from 1700 to 1730 was larger than the recent (post-1975) rise, indicating that changes of this magnitude and speed are not unprecedented in the past and so are not necessarily human-induced. Following this peak, the anomaly fell by 1.3°C in a couple of years; indeed in 1740 the anomaly was −2.53°C. Sudden switches of this type are not at all uncommon. We must presume that they are associated with either atmospheric circulation changes or SST changes in the North Atlantic, or other factors not yet explored or discovered.

Subsequently, temperatures varied up and down for about 155 years (1740–1895), on average being about one-quarter of a degree below the 1961–90 mean, but interspersed with years having temperatures as high as today's, as well as markedly much colder years than we now have.

There was great variability year by year over the full run of data, with swings of up to 2.5°C between consecutive years. (When anomalies covering larger areas are averaged, the extremes are reduced, since the averaging process lessens them. Over small areas, such as central England, the year-to-year differences can be high.) There were several cold spells, of over a decade in length, right up to the most recent one, starting in 1875. The Little Ice Age is deemed to have ended in the second half of the nineteenth century. (Since we are talking here about Northern Europe, not the globe, we can refer to the LIA quite legitimately.) A warming certainly does seem to

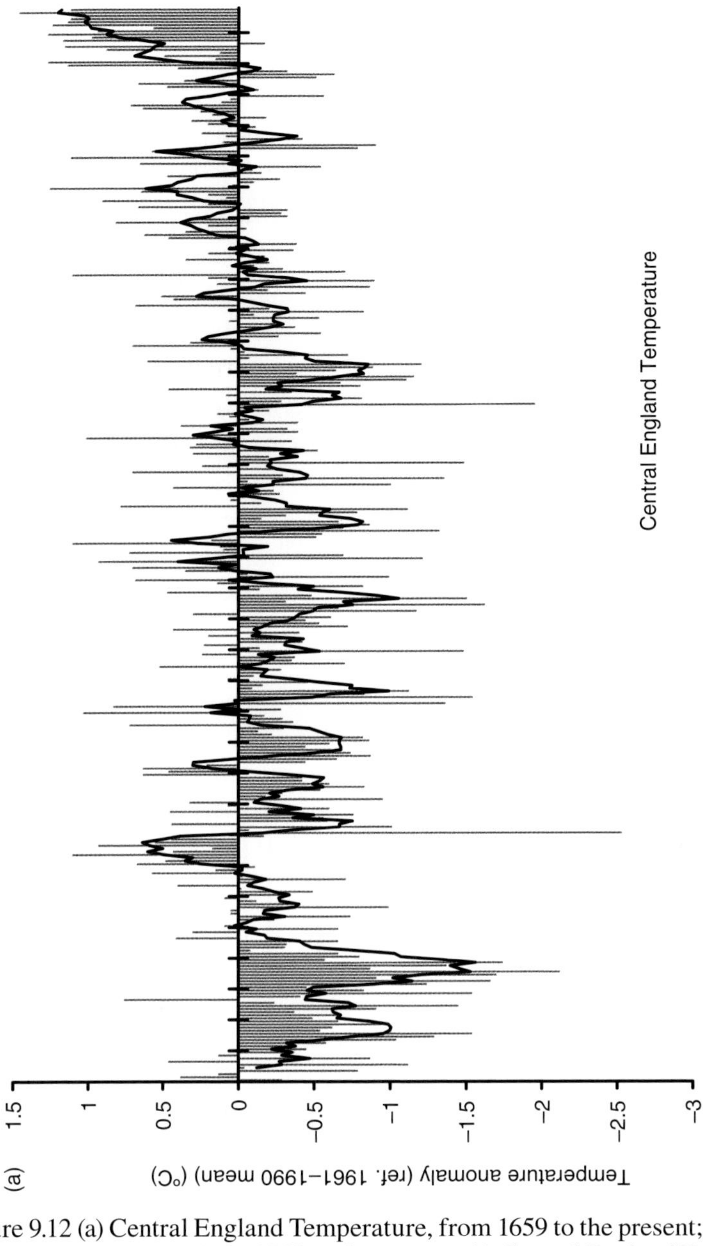

Figure 9.12 (a) Central England Temperature, from 1659 to the present; (b) Central England Temperature, from 1850 to the present; (c) Central England Temperature, from 1975 to the present. See text for discussion.

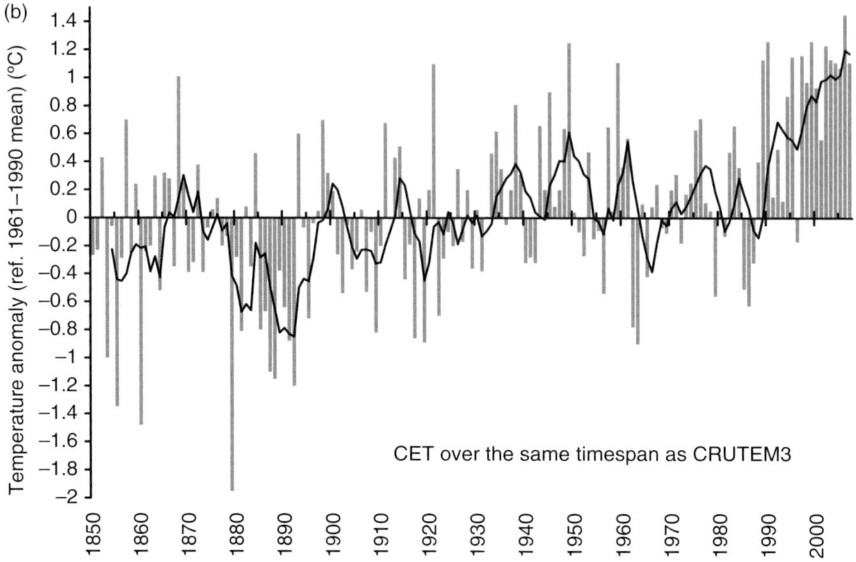

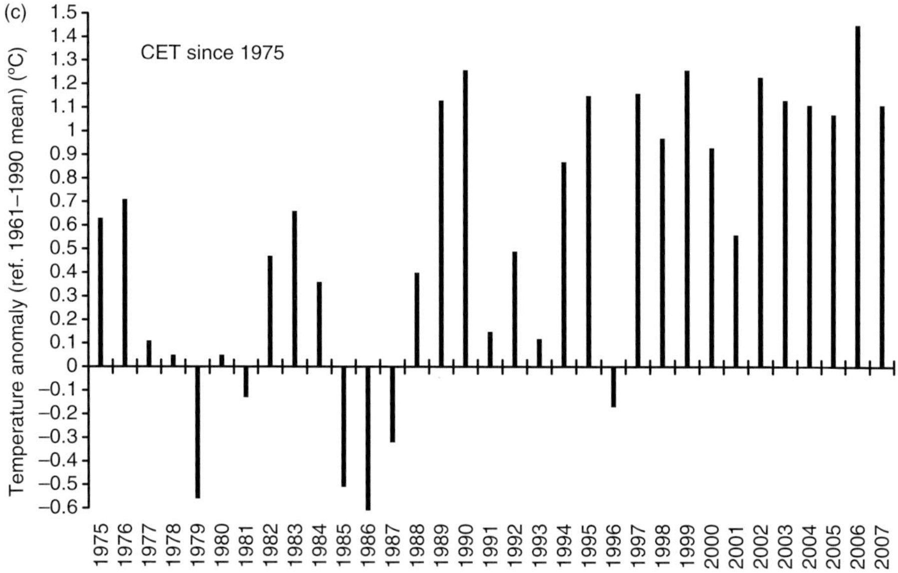

Figure 9.12 (*cont.*)

have started around 1890 that has continued up to the present day. The severity of the isolated (and occasionally grouped) much colder years reduced gradually after the cold period ended around 1700, suggesting a gradual move to less cold times (rather than to warmer times), the size and number of warm years remaining fairly constant after 1730, prior to which they were rare. The present warm years are hardly any warmer than many in the past; there are just more of them, closer together; the nearest equivalent period in the past would be the 1730s.

Temperatures from 1850 to 2008

Figure 9.12b is a plot of CET anomalies for the same period as the CRU/Had datasets for NH land (1850–2008) (Figure 9.6a), allowing a comparison to be made between them. Although there are many short-term differences, the two plots show similar long-term trends: a fall from 1865 to 1890, a rise until 1945–50, a fall to 1985, followed by the recent rise (which the next chart follows in more detail). A noticeable feature, even over this shorter period, continues to be the reduction in the amplitude of isolated cold years (a reduction that started in 1740, the coldest annual anomaly being –2.53°C in 1740).

To help understand these changes, Parker (2008) used techniques which he had developed for isolating the effects introduced by atmospheric circulation on temperature anomalies (see *Changes in atmospheric circulation* above). Applying these techniques to data back to the late nineteenth century, Parker found that atmospheric circulation explains much of the decadal and multidecadal variation of the CET in winter and some of the variability in spring. The spring and summer data match the SSTs of the mid North Atlantic, which changes with the *Atlantic Multidecadal Oscillation.*

However, the influence of the air source is limited by other factors such as changes in surface conditions where the air originates from and over which it passes. These factors include SST anomalies, snow cover, soil moisture changes, increases in greenhouse gas forcing and changes in cloud cover which, in turn, affect the radiation balance. Yet these various factors often develop *as a result of* atmospheric circulation patterns; for example, a long spell of mild south-westerly winds in winter produces higher than average SSTs around the UK, this then adding to the direct influence of the atmospheric circulation patterns. The overall effect of atmospheric circulation on seasonal CET will then exceed that directly resulting from each day's circulation. There is no simple explanation of the causes of CET changes; they are many and they interact. As ever, there is no simple one-reason solution. This logic, of course, applies to the whole globe, not just to England.

Temperatures from 1975 to 2008

Figure 9.12c is a plot of the last 30 years of CETs, showing swings above and below the mean until 1988, when there was a sudden shift to large positive anomalies; a state it has remained in until the present. There is no obvious trend, just a switch. The reasons for these changes involve the many, and interacting, factors listed in the previous section. Sometimes an anomaly might have a clear simple explanation. For instance, using the techniques described in the previous section, Parker (2008) found that the high CET in 2006 was due to high SSTs, and the low value in 1996 was due to atmospheric circulation. As in all graphs, the low values in 1991–3 were due to the eruption of Pinatubo.

Changes in maximum and minimum temperatures

A final matter that needs attention is that of changes in the *diurnal temperature range* (DTR) – the difference between daily maximum and minimum temperatures. The DTR is important because it contains information additional to the basic temperature, throwing some light on the possible reasons for some of the temperature changes. The work on DTR has progressed year by year since 1990, and it is interesting to follow the progress in chronological order.

Early investigations (the 1990s)

Karl et al. (1991) analysed monthly mean maximum and minimum temperatures over the USA, the USSR and China, and reported that most of the warming in these regions between 1950 and 1990 can be attributed to an increase in the minimum (mostly night-time) temperatures, which increased at around three times the rate of the maximum temperatures (mostly daytime), the latter showing little or no warming. This leads to a decrease in the DTR. The cause(s) of the observed changes were said to be uncertain, but with some evidence that changes in cloud cover played a direct role. These findings imply, the authors concluded, that either climate model projections are underestimating the rise of the daily minimum and overestimating the rise of the daily maximum, or that the observed warming in much of the NH landmass is significantly affected by factors unrelated to an enhanced anthropogenic-induced greenhouse effect. Further analysis a few years later by Karl et al. (1993), using gridded data covering 37% of the land surfaces, confirmed these findings. The IPCC Second Assessment Report (SAR; IPCC 1996) reported this work.

Easterling et al. (1997) extended the study to include 54% of land areas and lengthened the time series to 1993. They found that over the NH land surface, minimum temperatures increased by 0.18°C per decade, while the maxima increased by only 0.08°C, a slight reduction on the three-times ratio of Karl et al. (1991).

Later investigations (2001–2006)

The topic continues to have coverage in the IPCC's Third Assessment Report (TAR; IPCC 2001, p. 108), which says, as reported in the SAR and by Easterling et al. (1997), that the increase in temperature in recent decades has involved a faster rise in daily minimum than daily maximum in many continental regions. It goes on to say that while maximum temperatures have increased over most areas (with the exceptions of eastern Canada, the southern USA, eastern and southern Europe, southern China and southern South America) minimum temperatures have increased more almost everywhere (except eastern Canada and small areas of eastern Europe and the Middle East). As a result, the DTR decreased in most areas (except central Canada and parts of southern Africa, south-west Asia, Europe and some tropical Pacific islands). It is clear from this that there is great variability around the world with, for example, large reductions in the DTR over Venezuela and Columbia, with minimum temperatures increasing by up to 0.5°C. In northern China there is a similar rise in minimum temperatures, while in southern China a reducing DTR is due to a fall in the maximum temperature. So there is no universal pattern.

The TAR notes that the DTR is much affected by the urban heat island effect, with statistically significant differences between the DTR at rural and urban sites. This is, of course, not surprising because cities act as storage heaters by retaining heat gained during the day in the warmed concrete and bricks, releasing it slowly overnight, preventing the temperature falling as much as it would at a nearby rural site (assuming cloud conditions overhead are the same). But, as with temperature anomalies, unless there is a change in the city there should be no change in the DTR as measured in a city, so the DTR is unlikely to have been influenced by the UHI effect, unless the city grew in size (which it may have done, of course).

The TAR also observes that minimum temperatures in both hemispheres increased abruptly in the late 1970s, while at the same time there was also a change in the ENSO pattern (see Figure 9.3). The strongest changes in the DTR were in the NH winter, with a fall of 0.13°C per decade even at rural stations. The greatest observed warming (see Figure 9.6a) was also in the NH winter, suggesting that the two facts could be linked (Karl et al. 1991).

The TAR goes on to present charts illustrating the relationship between cloudiness and the DTR in Europe, the USA, the former USSR, mid-latitude Canada,

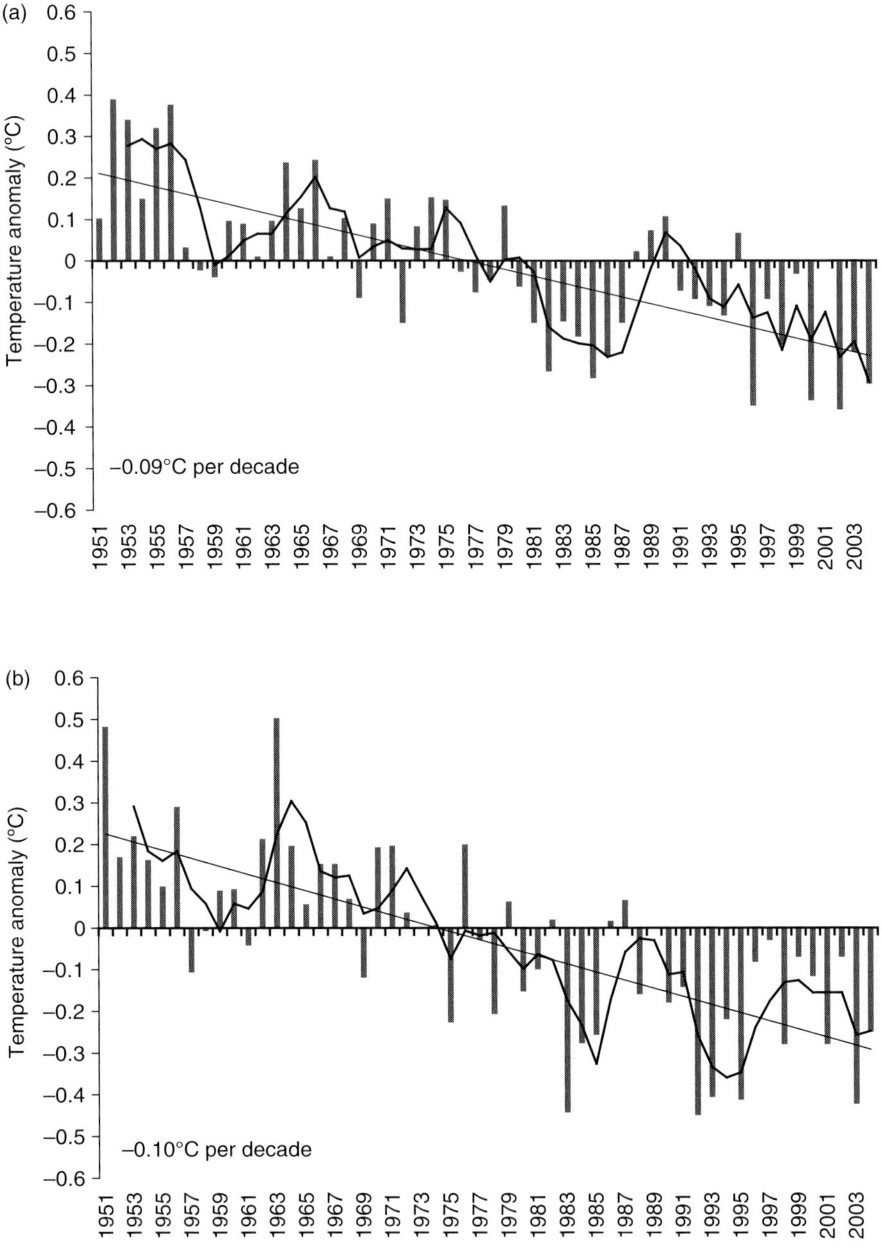

Figure 9.13 (a) Diurnal temperature range, Northern Hemisphere, Sep, Oct, Nov. (b) Diurnal temperature range, Northern Hemisphere, Dec, Jan, Feb. (c) Diurnal temperature range, Northern Hemisphere, Mar, Apr, May. (d) Diurnal temperature range, Northern Hemisphere, Jun, Jul, Aug. See text for discussion.

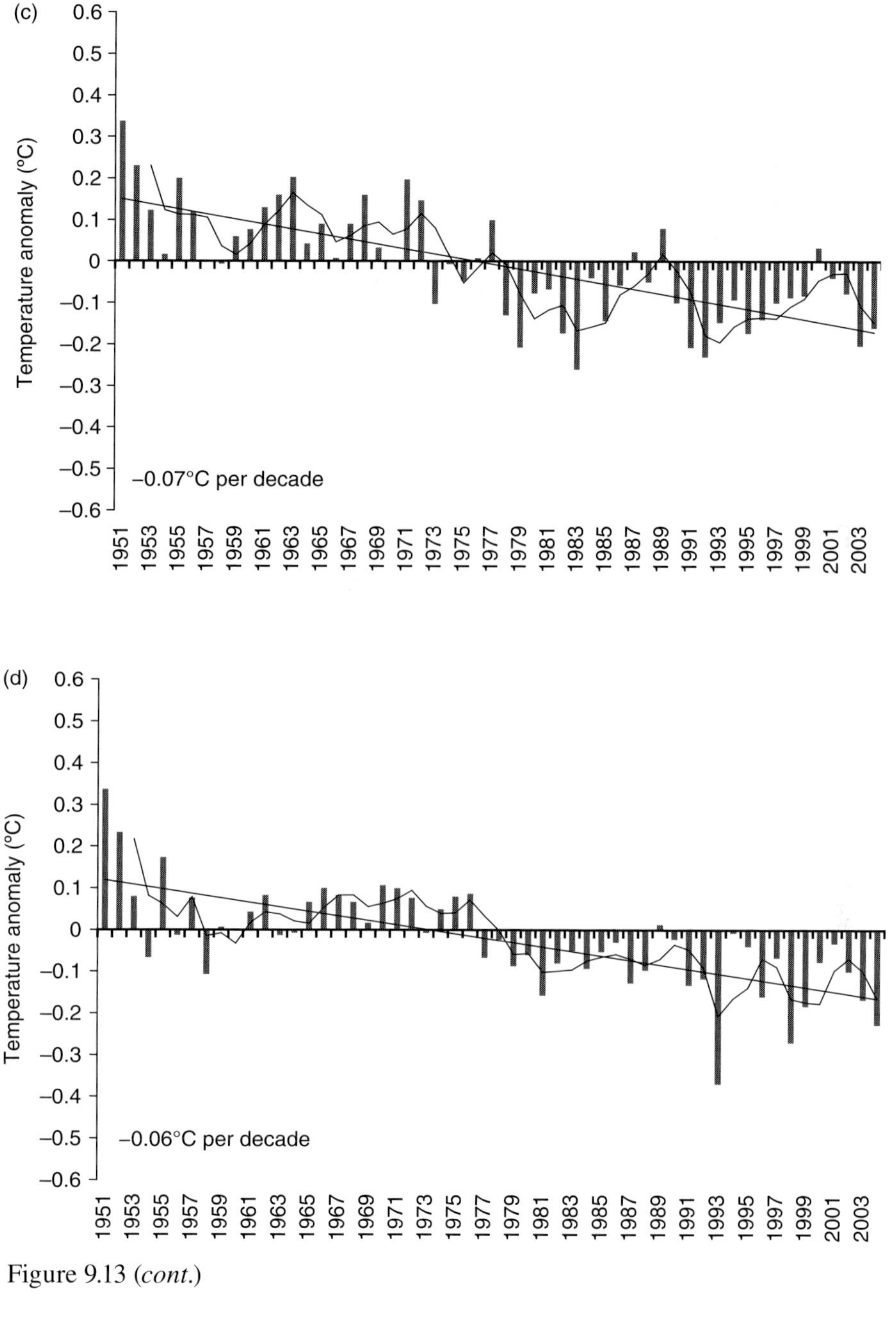

Figure 9.13 (*cont.*)

Australia and eastern China. In each region there was an increase in cloud cover over the twentieth century and generally a decrease in the DTR. In some instances, the IPCC report says, the correlation between annual cloud cover and annual DTR is 'remarkably strong', suggesting a distinct relationship between the two. This is not unlikely, for cloudy nights are warmer than clear nights because of the reduced heat losses through infrared radiation. An increase in cloud frequency will inevitably increase the number of warmer nights and this will reduce the DTR. This will apply in both rural and urban areas.

Since the mean air temperature is mostly calculated as the average of the maximum and minimum temperatures (see Chapter 8), a rise in the minimum temperature will cause the mean to rise, whether or not the maximum increases. Warming due to a reduced DTR caused by higher minimum temperatures is thus a sort of pseudo-warming because it is a matter of it being less cold at night rather than warmer by day. The question that arises, therefore, is how much of the warming in the NH over land in the winter is due to warmer nights, possibly resulting from increased amounts of cloud? It needs to be remembered that one of the major unknowns in climate models is the effect of clouds.

Recent investigations (2007 onwards)

The Fourth IPCC Assessment Report (AR4; IPCC 2007) introduces further new information regarding changes in DTR, drawing on the work of Vose et al. (2005), who reported studies that increase the coverage of maximum and minimum temperature measurements still further to 71% of the land surface of the globe. Russell Vose has kindly supplied me with datasets of the DTR and has agreed that I may use them here. They were produced using the *climate anomaly method* described in Chapter 8, with a base period of 1979–2004 (instead of 1961–90, as used by previous workers).

As Figures 9.13 and 9.14 show, there are considerable differences in the DTR between adjacent years and from season to season, while the hemispheres also have very distinct differences. The IPCC AR4 notes that the trends are highly variable from one region to another. This is another example of how one lumped (global) figure is not meaningful. It is impossible to sum up the changes in the DTR in one word, or one graph. By looking at the hemispheres and the seasons separately, however, much can be learnt; but it must be kept in mind that even these separate datasets still merge together different regions that will have very different DTRs. The only alternative is to look at regions separately, and that is not possible in such a brief review. Consequently eight graphs have been plotted to show the variations of the DTR for the two hemispheres and the four seasons separately.

Northern Hemisphere

The NH autumn (SON) (Figure 9.13a) along with that of the winter (DJF) (Figure 9.13b) show the largest changes in DTR. While the variability from year to year is large, there is a gradual fall overall, with a trend of −0.09 and −0.10°C per decade, respectively. Spring (MAM) (Figure 9.13c) and summer (JJA) (Figure 9.13d) follow a similar pattern to the autumn and winter, but with smaller amplitude swings and reduced trends of −0.07 and −0.06°C per decade, respectively. In all seasons there is a switch, around the mid-1970s, from above- to below-average DTR.

Southern Hemisphere

Because the SH seasons are opposite to those in the north, should we compare the same months of the year, or should we compare the same seasons of the same year? Indeed this question sums up the dilemma of merging everything into one, because by so doing we ignore these differences. As can be seen, however, there is more similarity between months (DJF, etc.) than between seasons (NH and SH spring, etc.).

The largest change in the SH's DTR is in the SH spring and summer (SON and DJF) (Figure 9.14 a, b); that is, in the same months as the largest NH changes – the summer being the larger with a trend of −0.07°C per decade, while the spring is −0.05°C per decade. As in the north, however, while there are considerable variations from year to year, there is a switch to negative anomalies in the mid-1970s, although this is slightly less pronounced than in the north.

The SH autumn (MAM) (Figure 9.14c) shows less trend (−0.04°C per decade), with positive and negative anomalies throughout the whole period, although with increasingly large negative anomalies for isolated individual years.

The SH winter (JJA) (Figure 9.14d) shows a completely different pattern, with almost no trend at all, or even with a slight positive upward movement of +0.004°C per decade, with positive and negative anomalies throughout.

All four SH seasons also show an increase in the DTR, with positive anomalies over the last three to six years. This is not observed in the NH. The two hemispheres are thus noticeably different.

Merging all eight graphs into one obscures the considerable seasonal and hemispheric differences and is thus not very informative. Nevertheless, to please those who like global figures for their simplicity, Figure 9.15 is included, showing both hemispheres and all seasons merged into one. It suggests that after 1979 the downward trend in the DTR stopped (it fell only 0.02°C per decade). Nevertheless the DTRs are all negative after 1979 (2001 excluded) showing that, while it is not falling, the DTR remains low. The lowest yearly value was also in 2000. The reason that the merged graph shows a reduced fall is because the SH DTRs have

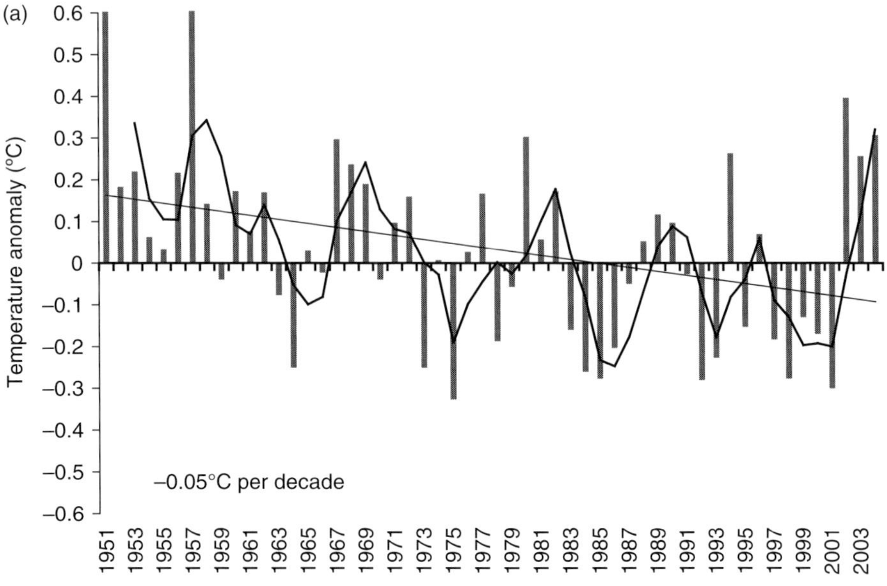

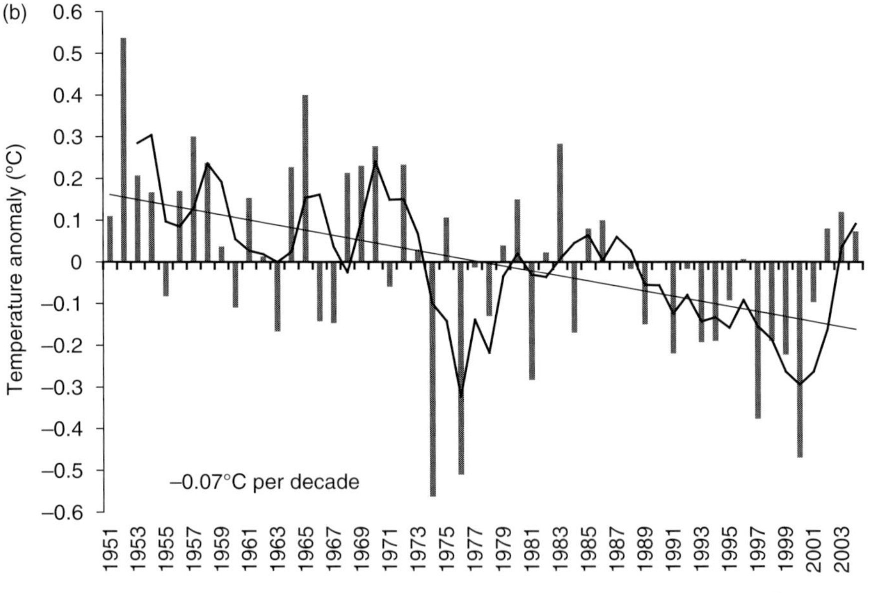

Figure 9.14 (a) Diurnal temperature range, Southern Hemisphere, Sep, Oct, Nov. (b) Diurnal temperature range, Southern Hemisphere, Dec, Jan, Feb. (c) Diurnal temperature range, Southern Hemisphere, Mar, Apr, May. (d) Diurnal temperature range, Southern Hemisphere, Jun, Jul, Aug. See text for discussion.

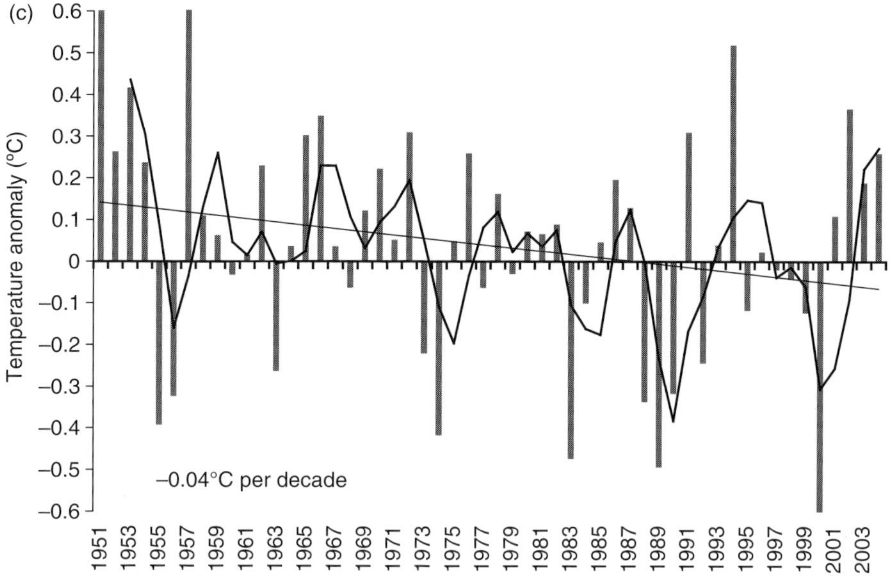

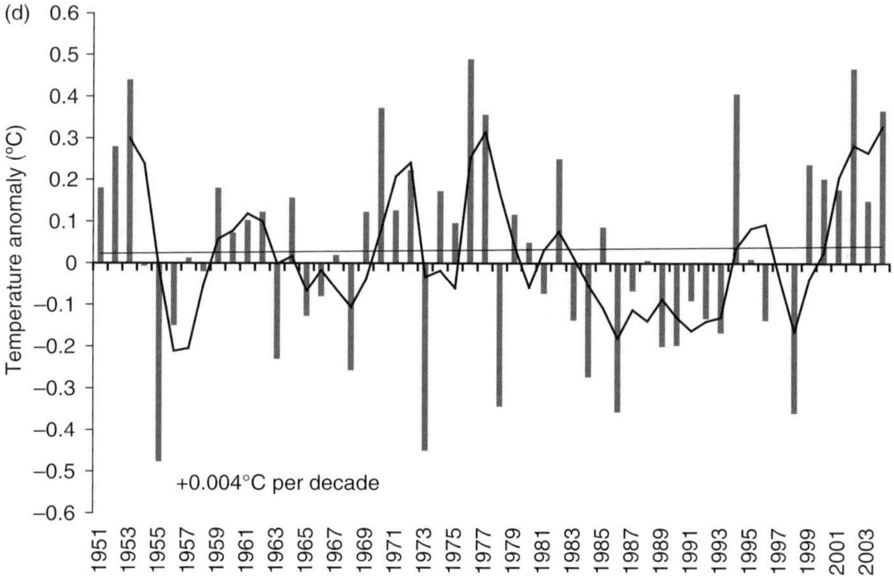

Figure 9.14 (*cont.*)

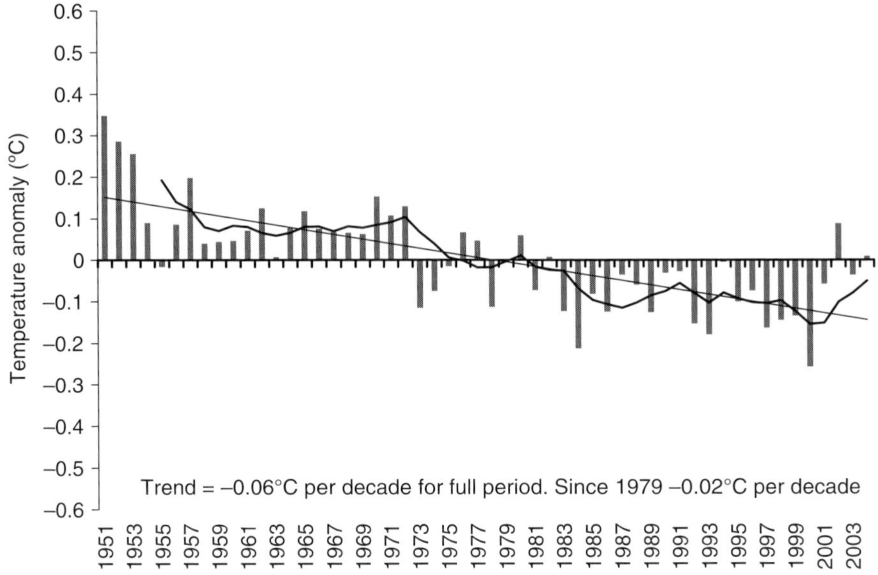

Figure 9.15 Diurnal temperature range for the globe for the full year.

all increased over the last three to six years, while this is not reflected in the NH. Additionally the SH has many positive anomalies after 1979, while the NH has just a few very small ones. The global figure is, therefore, misleading.

Causes of the changes in DTR

It is clear from the graphs that whatever causes changes in the DTR is variable in time and space over small time intervals, sometimes producing large switches of polarity between consecutive years, although there is a general overall pattern of reducing DTR in the NH in all seasons. The SH is much less affected, with only DJF and SON showing any obvious downward trend.

As noted by Karl et al. (1991, 1993) and in the IPCC AR4, increased cloud cover is seen as one of the most likely causes of the changes, although what has caused the cloud changes is another matter. Changes in atmospheric circulation are also seen as being potential influences, and these could have produced the changes in cloud cover. As with the causes of change in the CET, we are probably looking here again at complex interactions between several causes, all interacting at different times of the year and certainly different between hemispheres. The DTR is only known for the land because it needs the measurement of maximum and minimum temperatures at a fixed point, so we are ignorant of changes in the DTR over 71%

of the globe – indeed more, since there is little information from Antarctica. Our knowledge is consequently still quite limited.

The largest change in DTR, however, has been over the NH in winter (Figure 9.13a). This might be related to the fact that the greatest amount of warming has also been in the higher northern latitudes in winter (Figure 9.5). Yet again the complexity of the climate emerges. In Chapter 10 we look at measurements made up through the depth of the atmosphere and at what they show.

References

Allen, R., Lindesay, J. and Parker, D. (1996) *El Niño Southern Oscillation and Climate Variability*. CSIRO: Australia.

Ammann, C. M., Meehl, G. A., Washington, W. M. and Zender, C. S. (2003) A monthly and latitudinally varying volcanic forcing dataset in simulations of 20th century climate. *Geophysical Research Letters* **30**(12), 1657.

Bjerknes, J. (1969) Atmospheric teleconnections from the equatorial Pacific. *Monthly Weather Review* **97**, 163–72.

Brohan, P., Kennedy, J. J., Harris, I., Tett, S. F. B. and Jones, P. D. (2006) Uncertainty estimates in regional and global observed temperature changes: a new dataset from 1850. *Journal of Geophysical Research* **111**, D12106.

De Laat, A. T. J. and Maurellis, A. N. (2006) Evidence for influence of anthropogenic surface processes on lower tropospheric and surface temperature trends. *International Journal of Climatology* **26**, 897–913.

Easterling, D. R., Horton, B., Jones, P. D., Peterson, T. C., Karl, T. R., Parker, D. E., Salinger, M. J., Razuvayev, V., Plummer, N., Jamason, P. and Folland, C. K. (1997) Maximum and minimum temperature trends for the globe. *Science* **277**, 364–7.

IPCC (1996) *Climate Change 1995: The Science of Climate Change*. Contribution of Working Group I to the Second Assessment Report of the Intergovernmental Panel on Climate Change. Cambridge University Press.

IPCC (2001) *Climate Change 2001: The Scientific Basis*. Contribution of Working Group I to the Third Assessment Report of the Intergovernmental Panel on Climate Change. Cambridge University Press.

IPCC (2007) *Climate Change 2007: The Physical Science Basis*. Contribution of Working Group I to the Fourth Assessment Report of the Intergovernmental Panel on Climate Change. Cambridge University Press.

Jones, P. D. (2008) Historical climatology: a state of the art review. *Weather* **63**, 181–6.

Karl, T. R., Kukla, G., Razuvayev, V. N., Changery, M. J., Quayle, R. G., Heim, R. R., Easterling, D. R. and Cong Bin Fu (1991) Global warming: evidence for asymmetric diurnal temperature change. *Geophysical Research Letters* **18**, 2253–6.

Karl, T. R., Jones, P. D., Knight, R. W., Kukla, G., Plummer, N., Razuvayev, V., Gallo, K. P., Lindesay, J., Charlson, R. J. and Peterson, T. C. (1993) A new perspective on recent global warming: asymmetric trends of daily maximum and minimum temperature. *Bulletin of the American Meteorological Society* **74**(6), 1007–23.

Keeling, C. D. and Whorf, T. P. (2005) Atmospheric CO_2 records from sites in the SIO air sampling network. In: *Trends: A Compendium of Data of Global Change*. Carbon Dioxide Information Analysis Center, Oak Ridge National Laboratory, US Department of Energy: Oak Ridge, TN.

Kelly, P. M., Jones, P. D., Sear, C. B., Cherry, B. S. G. and Tavakol, R. K. (1982) Variations in surface air temperature. Part 2. Arctic regions, 1881–1980. *Monthly Weather Review* **10**, 71–83.

Lugina, K. M., Groisman, P. Ya., Vinnikov, K. Y., Koknaeva, V. V., Speranskaya, N. A. and Curtis, C. (2005) Monthly surface air temperature time series area-averaged over the 30-degree latitudinal belts of the globe, 1881–2004. In: *Trends: A Compendium of Data on Global Change*. Carbon Dioxide Information Analysis Center, Oak Ridge National Laboratory, US Department of Energy: Oak Ridge, TN. Available at http://cdiac.esd.ornl.gov/trends/temp/lugina/lugina.html.

Manley, G. (1953) The mean temperature of central England: 1698–1952. *Quarterly Journal of the Royal Meteorological Society* **79**, 242–61.

Manley, G. (1974) Central England Temperatures: monthly means 1659–1973. *Quarterly Journal of the Royal Meteorological Society* **100**, 389–405.

Mann, M. E. (2001) Climate during the past millennium. *Weather* **56**, 91–101.

Mann, M. E. (2004) On smoothing potentially nonstationary climate time series. *Geophysical Research Letters* **31**, L07214.

Mantua, N. J. Hare, S. R., Zhang, Y., Wallace, J. M. and Francis, R. C. (1997) A Pacific interdecadal climate oscillation with impacts on salmon production. *Bulletin of the American Meteorological Society* **78**, 1069–79.

McKitrick, R. and Michaels, P. J. (2004) A test of corrections for extraneous signals in gridded surface temperature data. *Climate Research* **26**(2), 159–73.

Parker, D. E. (2008) Anomalies of Central England Temperature classified by air source. *Journal of Climate* **22**(5), 1069–81.

Parker, D. E. and Horton, B. (2005) Uncertainties in central England temperature 1878–2003 and some improvements to the maximum and minimum series. *International Journal of Climatology* **25**, 1173–88.

Parker, D. E., Legg, T. P. and Folland, C. K. (1992) A new daily central England temperature series, 1772–1991. *International Journal of Climatology* **12**(4), 317–42.

Peterson, T. C., Connolley, W. M. and Fleck, J. (2008) The myth of the 1970s global cooling scientific consensus. *Bulletin of the American Meteorological Society* **89**, 1325–37.

Philander, S. G. H. (1990) *El Niño, La Niña and the Southern Oscillation*. Academic Press: New York.

Sato, M., Hansen, J. E., McCormick, M. P. and Pollack, J. B. (1993) Stratospheric aerosol optical depths, 1850–1990. *Journal of Geophysical Research* **98**, 22987–94.

Sexton, D. M. H., Parker, D. E. and Folland, C. K. (2003) *Natural and Human Influences on Central England Temperature*. Hadley Centre Technical Note 46. Met Office: Exeter, UK.

Slingo, J. M. (1998) The 1997/98 El Niño. *Weather* **53**, 274–81.

Smith, T. M. and Reynolds, R. W. (2005) A global merged land and sea surface temperature reconstruction based on historical observations (1880–1997). *Journal of Climate* **18**, 2021–36.

Thompson, D. J., Kennedy, J. J., Wallace, J. M. and Jones, P. D. (2008) A large discontinuity in the mid-twentieth century in observed global-mean surface temperature. *Nature* **453**, 646–649.

Vose, R. S., Easterling, D. R. and Gleason, B. (2005) Maximum and minimum temperature trends for the globe: an update through 2004. *Geophysical Research Letters* **32**, L23822.

Walker, G. T. and Bliss, E. W. (1932) World weather. *Memoirs of the Royal Meteorological Society* **4**, 53–84.

10

Temperature profiles through the atmosphere

So far, all of the measurements we have considered have been made at the surface, either over the land or over the sea. It is useful to investigate what temperature changes there have been up through the depth of the atmosphere.

Early measurements

The first upper air measurements were made around 1650, when Blaise Pascal and Florin Perrier carried barometers up a mountain, but because of the lack of suitable platforms to carry instruments further aloft, there was no further progress until about 1749, when Alexander Wilson made one of the first attempts using a kite, this being followed in 1784 by a balloon flight over Paris, giving the first observation of the temperature lapse rate. Balloons and kites developed alongside each other into the nineteenth century, when Joseph Louis Gay-Lussac made a balloon ascent to 23,000 feet (approx. 7,000 m), while in 1865 James Glaisher flew to 36,000 feet (approx. 11,000 m) in a coal-gas balloon, almost dying in the process. Kites continued to be used into the early twentieth century, tethers of piano wire allowing heights of 10,000 feet (approx. 3,000 m) to be reached, but this resulted in some disasters and a move was made to free-flying, unmanned balloons, much aided by the introduction of extensible rubber balloons in 1901, which allowed the first radiosondes to be developed (DuBois et al. 2002; Pettifer 2002).

Since the early 1950s, temperatures have been measured routinely through the depth of the atmosphere by sensors carried aloft by balloons (radiosondes, hereafter 'sondes'). Since the late 1970s it has also become possible to measure temperature profiles using *microwave sounders* aboard satellites. (These should not be confused with the satellite infrared measurements of surface skin temperature mentioned in Chapters 2 and 3 and discussed in Appendix D.) The important difference between sondes and sounders is that the former measure the temperature of the air directly, as they ascend through it, while sounders measure microwave radiation emitted from the atmosphere remotely, with the temperature being inferred indirectly using algorithms.

Radiosondes

The hardware

Sondes are disposable instrument packages lifted through the atmosphere by balloons, measuring profiles of temperature, relative humidity and barometric pressure. The first sondes were designed by the Soviet meteorologist Pavel Molchanov in 1928 and also independently in France and Germany at about the same time, while from around 1930 the UK operated sondes using sensors that varied in inductance. In 1931, Paul Duckert and Vilho Vaisala introduced capacitive sensors, which proved, eventually, to be the best type.

Modern sondes rise to an altitude of between 25 and 35 km (16–22 miles or 82,000–115,000 feet), which includes all of the troposphere and much of the stratosphere. As the balloon rises, readings are taken and transmitted back to a ground station. Since the instruments are rarely recovered in a reusable state, they are designed to operate for a few hours and then to be lost. Therefore, their cost must be low. Wind is also measured, either from the ground by following the sonde by radar, by means of a theodolite, by using navaid or, in the latest sondes, by including a GPS receiver aboard the sonde. Here we concentrate on the temperature sensors, their exposure, and what they show about temperature change over the last 60 years through the depth of the atmosphere. Details about humidity, altitude, position and the means of transmitting these data to the ground are peripheral to our main considerations and will not be discussed. For readers who would like to know more, there are several websites that give details of radio frequencies and on the history of sondes (see also Strangeways 2003).

In their brief upward journey, sondes encounter rain, snow, rime ice, sub-zero temperatures and very low atmospheric pressure, and they may pass through clouds of different types at several levels, all in the space of the hour or two of their flight, until the balloon bursts. This exposes sondes to a much more extreme environment than instruments experience at ground level, even in the Antarctic.

Unlike permanent instruments operated on the ground, sondes are not installed or maintained in the usual sense of the words, and are more akin to drifting buoys measuring SST, except that the latter have much longer lifetimes. Calibration also differs from the methods used at permanent ground stations because it is a once-and-for-all matter, but done in two stages. The manufacture will usually test every sensor at production to ensure that they meet their specifications, probably at only one or two points on the curves, and the user must then do tests just prior to launch to check them and to calibrate them against local instruments and conditions.

Today the key manufacturers are Vaisala (Finland), Sippican (USA), Graw (Germany) and Modem (France). (If I have omitted any, please let me know and I will put a note on my website at www.ianstrangeways.org.uk.) There are also manufacturers in China, India, the former USSR and Messei (Japan). Information on how many sondes are made by these various manufacturers is not available, this sort of detail being confidential or simply unknown. In the past there were innumerable makes; for example, Plessey manufactured such instruments in the UK and there were many made by (or for) the National Meteorological Services, including the UK Met Office, for their own use.

Since designs were based on the technology available at the time of development, they have changed dramatically in the 60 years of their existence. Figure 10.1 illustrates an early sonde, which was largely mechanical, while Figure 10.2 shows a modern design, which is entirely microelectronic. Initially there were many designs, all having different accuracies and errors, but over time there has been a convergence, with a move towards miniaturisation, reduced weight, lower cost, and improved sensors and electronics. But significant technical differences, and errors, still remain. For example some sondes use thermistors, others use capacitive temperature sensors. Humidity and altitude sensors also vary, as do battery types. Military sondes are different yet again.

The improvement in design over the 60 years of sonde use, while improving accuracy, also introduces major problems for climate researchers looking for long-term trends, because every change, by every manufacturer, introduces another change in errors and accuracy. This is not unlike the problems that had to be addressed when producing datasets from both land air temperature screens with mercury thermometers and SST measurements from many different sources. Recall that none of these instruments were developed or operated for the purposes of climate research, nor are they still.

Sonde temperature sensors

Temperature sensors used in sondes need to respond rapidly and should have a time constant of around one second to ensure that errors from thermal lag are less than 0.1°C while rising through any 1 km layer of atmosphere (Figure 10.2, right). This calls for a small device with maximum surface-area-to-volume ratio. While radiation screens perform adequately at the high pressures of lower altitudes, they present considerable problems as the pressure falls, their surfaces becoming much warmer or cooler than the atmospheric temperature, depending on the terrestrial and solar radiation conditions. Although a screen made of two concentric polished

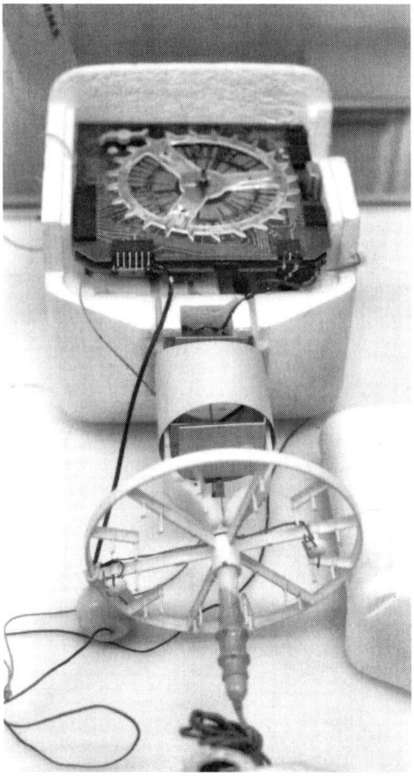

Figure 10.1 In this early radiosonde, the propeller turns as the sonde ascends through the atmosphere rotating the switch above it, routing the sensors sequentially to the radio transmitter. The sonde was kindly made available by Jane Insley at the Science Museum, London.

aluminium cylinders with a spacing of 1 or 2 cm might be acceptable, measurements have been shown to be more accurate when the sensor is exposed directly without any protection, although this only applies to the smaller modern sensors (as shown in Figure 10.2).

To minimise the radiation errors that result from having no screen, the sensor must have a surface that absorbs as little solar and IR radiation as possible, and this can best be achieved by depositing a very thin metallic film on it. Many of the white paints used in the past have high absorption in the IR spectrum, even though they reflect the visible wavelengths (appearances can be deceptive, and one wonders about the white paints used on Stevenson screens). A drawback of having no protection is that the sensors are exposed to rain, cloud droplets and ice formation. The sensors must shed these if large errors are to be avoided. Sensors are best exposed above the sonde, to avoid heating, or cooling, by the package. This is achieved by mounting the sensor on a short arm just higher and to the side of the

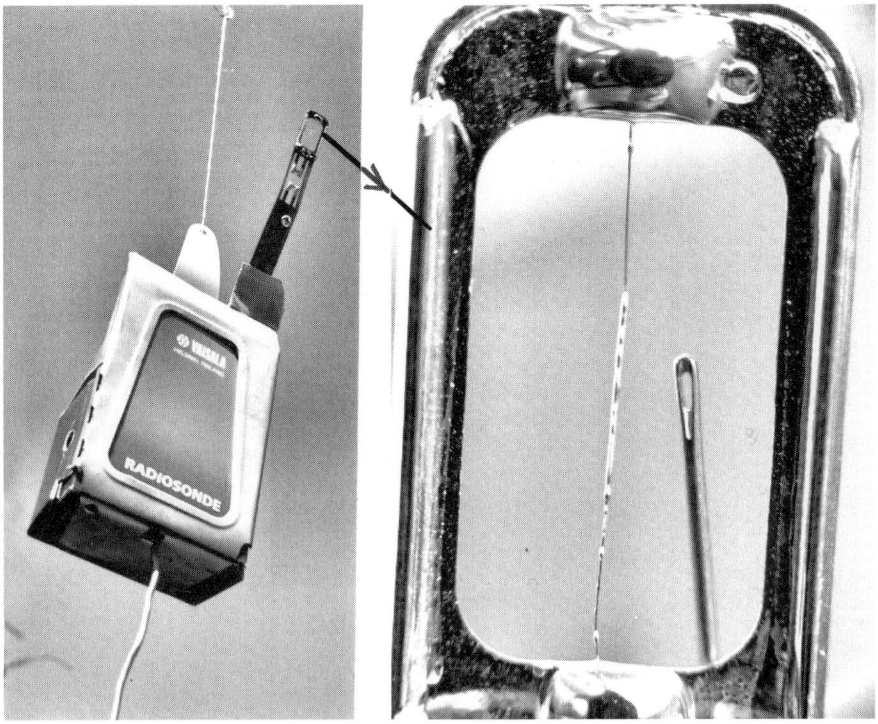

Figure 10.2 The temperature sensor in this modern sonde is at the top of the arm. A close-up of it to the *right* shows its small size (to minimise radiation exchange and ensure a rapid response). Beneath the temperature sensor is the humidity sensor (not shown in close-up). On the latest design, a GPS antenna is also affixed to the top of the sonde. More details are given in the text. The sonde was kindly loaned by Vaisala Ltd.

package (Figure 10.2, left). The sensor also needs to be robust enough to survive rough handling during launch (Figure 10.3), often done under more difficult field conditions than shown in the photographs.

In the past, nearly every type of temperature sensor has been used for sondes, even bimetallic strips (despite their slow response and the problems of shielding them). But today the preferred sensor is a thermistor or thermal capacitor, each of which can be made very small and can be suitably protected from radiation by a thin reflective film. Thermal capacitors were developed especially for sondes and can take various forms, but that illustrated in Figure 10.2 is of the more recent thin-wire type made of two fine platinum wires separated by a glass-ceramic dielectric, the value of which changes with temperature, causing a change of capacitance between the wires. The whole is sealed in a glass coating, which is covered in a reflective aluminium film.

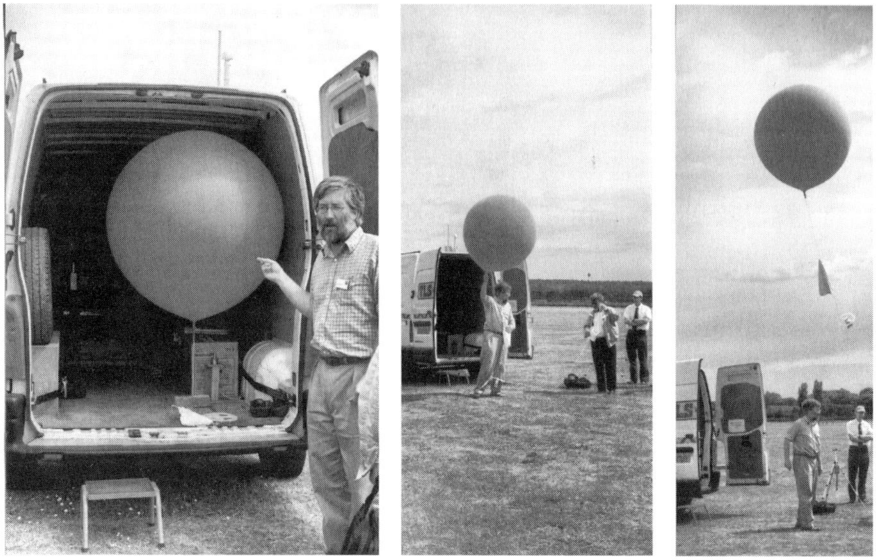

Figure 10.3 Launch of a radiosonde at a meeting of the Royal Meteorological Society Special Interest Group on Meteorological Observing Systems.

Accuracy, calibration and errors

Each make of radiosonde has its own range of errors, although it would be inappropriate to list them here for all makes. For those who need such detail, the WMO handbook (1996a) includes tables comparing the various makes of sonde, old and new, currently in use. The WMO has also carried out extensive intercomparisons of sondes (WMO 1987, 1991, 1996b). A summary of the errors that all sondes are subject to, and of the performance of just the latest sondes, will suffice here.

General sources of error that affect all sonde temperature sensors are errors in the factory production/calibration process, errors in pre-flight check readings (see below) and problems arising from interpolating between readings when transmission is intermittent due to poor radio reception. In addition, IR heating of the sensor occurs day and night, from the atmosphere above and from the atmosphere and the ground below, with the latter being the greater. As the sonde rises, the IR flux changes, and it also varies from flight to flight depending on the cloud present and the temperature of the atmosphere and of the ground. But the sensor also emits IR itself, and to obtain the correct reading, absorption and radiation must be in balance. Under some conditions the sensor can cool by IR radiation well below the air temperature. The balance varies with pressure and therefore with height, being most skewed at the lower pressures, but the balance is also dependent on the actual air temperature, so the situation is complex and variable. To complicate matters further, latitude is also a factor, and because conditions vary so much from flight to flight, the errors of individual flights are difficult to correct for. The older sondes, with white rod thermistor sensors, are more prone to these errors. In the older designs, thermal lag is a further source of error, but again the newer models are free of this.

Rate of ascent must also be taken into account, since this affects ventilation. At sunrise and sunset, corrections must take into account sonde height because errors are very sensitive to solar elevation. The presence of cloud must also be allowed for in correcting for solar heating. The albedo of the ground surface is an additional and important factor, with low cloud and snow reflecting a large proportion of the incident solar radiation, while in contrast the sea has a low albedo. Sondes also swing and rotate beneath the balloon and this can allow air warmed by the sonde body to contact the temperature sensor, this effect increasing rapidly in the low pressure of the stratosphere. The motion of the sonde beneath the balloon also varies the orientation of the sensors to the Sun, and thus the amount of radiation incident on it.

If water or ice accumulates on the temperature sensor, further errors are caused due to cooling (wet bulb effect) after the sonde has left the rain or cloud belt and the relative humidity (RH) has fallen below 100%. Ice forming over the metallised reflective skin acts as a 'black body' and leads to radiative cooling. If water, formed on the sensor at low altitude, freezes, the released latent heat will warm the sensor towards 0°C. Conversely, if an ice-coated sensor moves into warmer air, the reading will stay at freezing point until all the ice has melted.

To minimise errors, sonde readings must be checked manually *immediately* prior to launch. This applies to all sensors (temperature, humidity and pressure). Any error in these readings is included in the sonde reading, and errors in manual observations may not be small. The sonde's accuracy thus depends very much on the quality of the instruments used in pre-flight tests and on the observer's skill in reading them. Although pre-launch checks give confirmation at one point on the sensor's response curve, the performance at other points cannot be checked, and the manufacturer's calibration has to be relied on. Reliability varies from make to make.

It will be apparent, therefore, that no simple universal statement about temperature accuracy is possible, given the wide range of factors that affect it and also the wide range of instruments manufactured. However, as a rough guide, the best of the modern sondes measure temperature to a resolution of 0.1°C, to an accuracy of ±0.2°C, with a response time of 0.2 s at ground level, slowing to 1.0 s at 10 hPa. This includes the variability introduced by virtue of the fact that every flight is different.

While I have drawn attention in earlier chapters to the many shortcomings of Stevenson screens and mercury thermometers, the situation with radiosondes is even less satisfactory. Only one observation is made daily (perhaps two at some stations). Although wooden screens on the land are not all the same, and can deteriorate, they have usually remained the same at any one site for decades, whereas sondes have changed rapidly as technical improvements and new developments occurred, and there has been little international consistency in the past. We thus have a much less substantial and shorter set of measurements in the case of sondes.

Perhaps one way in which the sonde data might be thought of as better is that the measurements are not affected by the surface conditions (land-use, UHI). But, even here, one could argue that as the sonde rises it may ascend within a thermal, adding its speed to that of the uprising air, right up to the stratosphere, thereby being greatly affected by surface conditions which are carried aloft by the air itself; the sonde rising in a rising bubble, especially in the tropics, where large cumulus clouds dominate. A further problem is that in the one ascent of the day the sonde may pass through dense, deep, stratus cloud one day but through clear sky the next. Can one really detect a small long-term trend in amongst all of this great variability, not to mention the periodic changes made over time to the sonde designs?

Review of performance and of corrections

Not surprisingly, from the foregoing review, the IPCC (2007) state that uncertainties about long-term changes in the radiosonde temperature records are substantial. Changes in instrumentation and procedures permeate both the sonde and the satellite records, concealing the small long-term trends. Consequently it is difficult to make any reliable judgements as to which, if any, of the many independently derived estimates is closest to the true change in climate (Karl et al. 2006).

Much effort has been made to assess and improve the quality of the sonde temperature record, for example to reduce the effects arising from instrument and procedural changes during the seven decades of sonde operation (Free and Seidel 2005; Sherwood et al. 2005; Karl et al. 2006). However, a comparison of several of these methods shows that they gave dissimilar results when applied to one common set of sonde data (Free et al. 2002). A comparison of five sonde datasets agreed well enough over the short term, for example in responding to ENSO cycles and volcanic eruptions, but over the long term they were inconsistent. It is the latter that matter in the context of climate change.

Several approaches have been used to generate improved datasets from the old records, with some of the errors corrected for, Lanzante et al. (2003a, 2003b) using 87 carefully selected stations and applying subjectively derived bias adjustments – although subjectivity is not a very sound basis. Another new record, HadAT2 (successor to HadRT), compares neighbouring stations, much like the methods used when correcting surface data. A third approach (Haimberger 2005) uses model-based reanalyses to reduce inhomogeneities in the data. Rather than adjusting the data, Angell (2003) attempted to reduce data-quality problems by removing several tropical stations, known to be unreliable, from the network (Santer et al. 2008), but this does not deal with the problems inherent in even the best-operated and best-designed instruments. Of these approaches, the one represented by HadAT2 appears to be the soundest.

Despite these efforts, significant problems remain. Sherwood et al. (2005) found substantial changes in the diurnal cycle in unadjusted data, probably due to improved sensors and radiation protection over time. Relative to night-time values, they found a daytime warming prior to 1971 and then a daytime cooling from 1979 to 1997 that are probably false and instrument-induced. It is estimated that there has been a spurious reduction in sonde temperature records since 1978, throughout the atmosphere, of about 0.1°C per decade globally, this being greatest in the tropics (0.16°C per decade) and least in the NH extratropics (0.04°C per decade). Randel and Wu (2006) also detected problems suggesting that negative biases are not limited to daytime observations, although a few stations may have positive biases (Christy and Spencer 2005).

In addition to the above reservations and caveats, sonde datasets are also largely limited to NH land areas, with the coverage being poor over the tropics and the SH generally, partly because there is less land in the SH. Global estimates of temperature trends based on sondes are thus neither highly reliable nor globally representative (Agudelo and Curry 2004).

Sondes do a very good job for meteorology and weather forecasting, the purposes for which they were originally designed, are made, and are still operated, but we should beware of placing too much reliance on them for climatology. But do sounders do any better?

Sounders

Principles

An alternative method of measuring temperature profiles through the atmosphere is by means of microwave sounders. Natural thermal radiation from the atmosphere in the microwave band originates from oxygen, water vapour, liquid water (cloud droplets) and carbon dioxide at very specific frequencies, the intensity of the radiation depending on the temperature and quantity of the species. Since the distribution of oxygen with altitude is well known, and is constant over time, its radiation contains information on temperature. Temperatures at different heights can be obtained by making measurements at different frequencies.

Sounder hardware

The first microwave sounder used for measuring temperature, the *Microwave Sounding Unit* (MSU), became operational in late 1978 on the NOAA's TIROS-N satellite. It had a ground resolution of 50 km and measured a total of 15 frequencies, giving profiles of temperature from the ground to 45 km (3 hPa). This evolved into the *Advanced Microwave Sounder Unit* (AMSU), the first being aboard the NOAA-15 satellite launched in 1998, followed by NOAA-16 (2000), NOAA-17 (2002) and NOAA-18 (2005). These satellites carry two sounders, AMSU-A which measures temperature and AMSU-B which senses water vapour, the latter being replaced in 2005 by a similar instrument known as the *Microwave Humidity Sensor* (MHS). Versions of the AMSU-A also fly on NASA's 'Aqua' Earth observation satellite and EUMETSAT's MetOp series of satellites. The next generation in the series is known as the *Advanced Technology Microwave Sounder* (ATMS), due to become operative some time between 2009 and 2010.

These instruments are all operated on satellites in polar, sun-synchronous, orbits, resulting in them crossing the equator at the same two local solar times each day. The sounders scan continuously *across-track* ('whisk broom' mode) as the satellite moves forward, making 30 readings in 3.3° steps from −48° to +48°, followed by a reading of the on-board 'warm' calibration target and of 'cold' space (as an additional calibration point), returning to its starting position to repeat the cycle, the whole of which takes eight seconds. During this time the satellite has moved forward 45 km further along its path. About 16 orbits are made every day, allowing the globe to be observed in either daylight or darkness, sometimes both. The polar regions are passed over about every 100 minutes, but any particular point on the equator much less often. The 'spot size' of the observation is 45 km directly beneath the satellite, becoming elongated towards the edges of the scan, or 'swath'.

The AMSU-A sounder has 15 channels, with channels 5–8 measuring from 53.6 to 55.5 GHz to sense tropospheric temperature at various depths, and channels 9–14 measuring stratospheric temperature in the spectrum from 54.4 to 57.3 GHz.

Brightness temperature and algorithms

A microwave radiometer does not measure actual air temperature, but what is called *brightness temperature* (see also Appendix D). This is a function involving the radiance of the target (oxygen molecules in this case), the wavelength of its emissions and Boltzmann's constant (which relates energy at the particle level with temperature observed at the bulk level). There is, therefore, no direct and simple equivalence between the measured microwave radiation intensity and the temperature of the object emitting it. Models or *algorithms* have to be developed to convert from the radiance measurement to degrees of temperature. These algorithms can be quite complex.

Obtaining temperature profiles from sounder measurements

Sounders receive an intricate mix of microwave radiation emerging from the top of the atmosphere. The algorithms used to convert the measurements to temperature involve complex calculations known as *radiative transfer equations* (RTEs). (Radiative transfer refers to the physical phenomenon of energy transfer in the form of electromagnetic radiation.) The propagation of radiation through a medium (the atmosphere in this case) is affected by absorption, emission and scattering processes. The equation of radiative transfer describes these interactions mathematically, and involves the Planck function. The equation is in two parts: the first representing the radiation originating from the surface, the second radiation from the atmosphere. The latter incorporates a *weighting function*, which gives profile information, the function peaking at different altitudes, allowing temperature soundings to be made by selecting wavelengths that correspond to radiation originating (largely) from specific levels. The solution of the RTE is complex, in part because the weighting functions overlap, making it difficult to sense the radiation from a precisely specified level. Comparisons with radiosondes have helped check the accuracy of satellite retrievals, but sondes have their own errors and there are errors caused by differences both in the timing of the measurements and in differences in location.

To obtain a profile of temperature through the atmosphere, the thermal emissions of oxygen are measured by the microwave sounder at around 54/55 GHz, giving an indication of temperature in the troposphere. Measurements at around 57 GHz are sensitive to the temperature in the lower stratosphere. These measurements can be combined to give an indication of the average temperature of the lower atmosphere, weighted mostly in the region of 1.5–5 km altitude.

Channel 2 of the MSU measures a thick layer of atmosphere, about 75–80% of the signal originating in the troposphere, 15% from the lower stratosphere and 5–10% from the surface. Channel 4 is sensitive mostly to the lower stratosphere. The vertical resolution is thus very coarse and, coupled with the large footprint of 45 km, the volume of air measured is considerable.

Additional problems arise when data from all of the satellites are merged to produce a continuous record, the process having to allow for differences in sensor calibration between satellites and differences in the on-board calibration targets. Changes in orbit (due to the slight drag of the very thin remaining atmosphere that satellites experience in low Earth orbit) alter the satellite's orbit slowly and thus change its altitude and the time of day it passes over any one location. The measurements need to be made at the same time at every location to minimise the diurnal changes in temperature.

Assessment of sounder performance

Independent groups have used different methods to correct for the errors just described (Thorne 2008) and have found global warming of the troposphere to be anywhere between 0.04°C and 0.2°C per decade for the period 1979–2000. This very large spread is due to differences in the calibration of the sensors, the method of merging the data from different satellites and the adjustment for orbital changes, diurnal cycle changes and variations in the calibration target temperatures (e.g. Mears et al. 2003; Christy and Norris 2004; Grody et al. 2004; Fu and Johanson 2005; Vinnikov et al. 2006).

Comparison of sonde and sounder datasets (Christy and Norris 2004) show considerable scatter and differences between them. In view of the direct nature of the sonde data compared with the indirect and uncertain accuracy of the satellite data, only sonde data will be examined here.

Temperature profiles from sonde data

There have been, and still are, uncertainties about whether the surface and upper air measurements agree or not, with suggestions by some that there has been no warming in the troposphere equivalent to that at the surface (and that, therefore, either the models are wrong for predicting it, or the surface measurements are wrong – suggesting that warming has not occurred). The discrepancy is particularly noticeable in the tropics. But what do the measurements actually show?

Causes of tropospheric warming

If the surface warms, we should not be surprised if tropospheric temperatures follow the same pattern as those at the surface. The Greek word *tropo* means turning over, or mixing, and if the air at the surface warms, the air higher up is also likely to warm (as explained in Chapter 1) through convection and turbulence. By these processes, heat is transported throughout the troposphere, up as far as the tropopause, along with the transfer of latent heat through the condensation of water vapour into clouds. So if the surface warms, the troposphere should too, although to what extent and to what altitude is dependent on conditions and on the causes of warming at the surface.

Again we are looking here at changes in temperature, not at absolute temperatures. Clearly the absolute temperatures will be much lower the further from the surface the measurement is made, due to the lapse rate.

There is also the matter of warming of the atmosphere by IR radiation exchange between the ground and the atmosphere. Clouds reduce the amount of IR radiation reaching the troposphere and the same clouds also keep the surface warmer by preventing the loss of the IR radiation (by producing less cold nights and so a reduced DTR). Also plentiful water vapour may absorb all the IR near the surface.

Since much careful effort has been put into attempting to overcome the numerous problems that beset sonde measurements, it is worth looking at what the adjusted datasets show, and for this purpose I use the HadAT2 records produced by the Hadley Centre, who have kindly agreed to me using the datasets for illustration.

The IPCC AR4 (2007) shows graphs of global temperature anomalies based on HadAT2 measured by sondes over the period 1958–2005 for the lower and the mid-to-upper troposphere, as well as for the stratosphere. The IPCC graphs also include curves based on three sounder datasets, from the University of Alabama in Huntsville (UAH), from remote-sensing systems (RSS), and a dataset by Vinnikov et al. (2006). Again we have the problem that globally merged figures, north and south for all seasons, with the troposphere subdivided into only two coarse layers, are not the best of indicators, since they hide much important detail.

The HadAT2 datasets, however, include a great deal more than the one global figure, including the temperature anomaly for each month, in each latitudinal band of 5° width, for heights of 850, 700, 500, 300, 200, 150, 100, 50 and 30 hPa pressures.

Global temperature trends through the full depth of the troposphere

It is not possible (here) to show what happened in each season, at each altitude and at each latitude. I have, instead, plotted the datasets that merge all of the latitudes north and south and all of the altitudes up to the tropopause, giving an overall global picture of changes in the troposphere (Figure 10.4). I have done this despite not liking gross global values, but it is unavoidable and has some usefulness in this case, because if it is laid alongside Figure 9.6c (of global land air temperatures at the surface), Figure 10.4 shows an overall agreement with it; even the 1998 El Niño year and the Pinatubo years of 1991–3 standing out very prominently. There is also the same fall of temperature up to 1976, followed by a very sudden switch to large positive anomalies after 1980, reflecting similar tendencies in the surface record. This would seem to suggest that the sonde measurements agree reasonably well with the surface measurements.

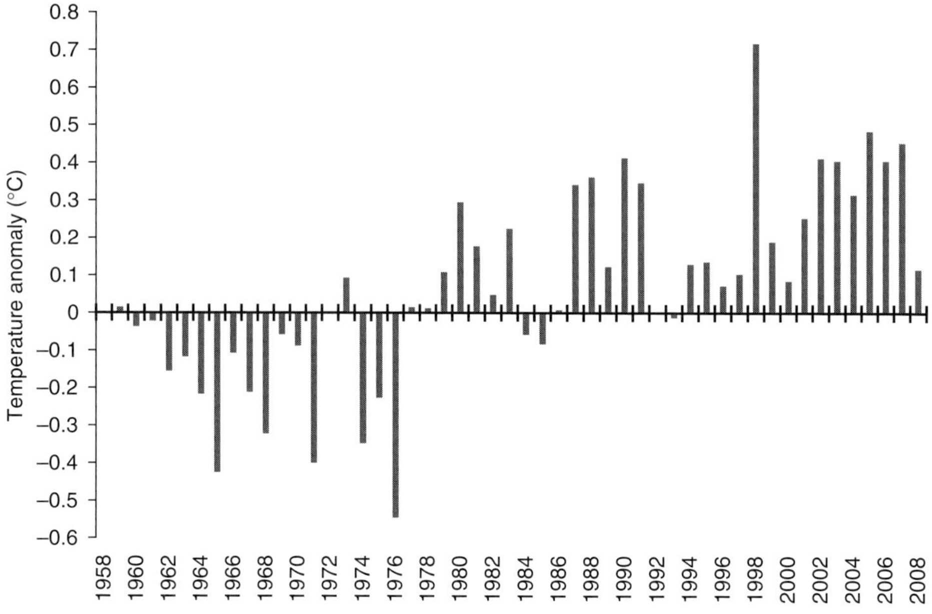

Figure 10.4 Radiosonde record averaged for the whole globe up to the tropopause. See text for discussion.

And yet, despite this, Thorne et al. (2005) sum up the situation thus: 'We can no longer conclude whether globally the troposphere is cooling or warming relative to the surface'. This statement, incidentally, includes the satellite sounder data.

Temperature trends in the mid-troposphere at 67.5°N in winter

I have also plotted the anomalies for December, January and February separately in high northern latitudes (67.5°N), because this is when, and where, most of the recent surface warming has occurred (Figure 9.5). Figure 10.5a, b, c shows the temperature anomalies at an altitude of 700 hPa (about 10,000 feet (3,000 m) or mid-troposphere). Not surprisingly, at these monthly time intervals, the year-to-year variability is much greater than in a larger merged dataset, with swings of several degrees from one year to the next. But what trends emerge? Do they reflect those trends at the surface, or the global sonde records?

A visual comparison between Figures 10.4 and 10.5a shows immediately that the December anomalies at 67.5°N and about 10,000 feet (3,000 m) altitude are very different from the global anomalies for the full depth of the troposphere. There is no clear-cut switch – most apparent in the global chart around 1978 – from falling large negative anomalies to rising positive anomalies, with positive and negative anomalies coexisting throughout the period, although there is a pattern of declining positive anomalies up to 1978 followed by rising positive anomalies thereafter, with an overall upward trend of 0.12°C per decade. January (Figure 10.5b) shows a slight upward trend (0.10°C per decade), with the same fairly static response after 2000. February (Figure 10.5c) is different yet again, with a zero trend. By combining the anomalies of these three months to give a 'winter' chart (Figure 10.5d) the extremes are smoothed out, resulting in lower swings. The overall upward trend is 0.07°C per decade. A slight rise occurs after 2000, which subsides over the last three years. However, none of these latter years are warmer than many of the foregoing 40 years; there are no broken records amongst them – indeed this is true of each of the three months.

There is no sign, therefore, in the northern, high-latitude, mid-troposphere winter, of the sudden large warming seen over the land surface beneath. The IPCC AR4 (2007, p. 271) comment on this specifically, saying

> The global trends since 1979 disguise many regional differences. In particular, in winter much larger temperature trends are present at the surface over the northern continents than at higher levels (Karl et al. 2006).

On the same theme, Vinnikov et al. (2006) say

> At middle and high latitudes in the Northern Hemisphere, the zonally averaged temperature at the surface increased faster than in the troposphere, while at low latitudes of both hemispheres the temperature increased more slowly at the surface than in the troposphere.

The lack of an equivalent rise in the troposphere in the high northern latitudes in winter to match that at the surface points to the possibility that increased cloud cover may have caused the extra surface warming. Such warming would not be reflected in the troposphere above, since the effect would be restricted to under the clouds, not above them; indeed the clouds would insulate the atmosphere above from the warming beneath. This possibility is supported by the finding that the DTR has fallen, particularly in the high northern latitudes in winter, and that this has a high correlation with increased cloud cover. If this is the case, then the anomalously higher warming in the NH lands in winter is due to clouds – to warmer nights rather than to warmer days.

Long-term temperature trends in the troposphere are still not fully explored, with differences between latitudes, seasons, altitudes and hemispheres, and with considerable uncertainty over the tropics. There is still much to be learnt.

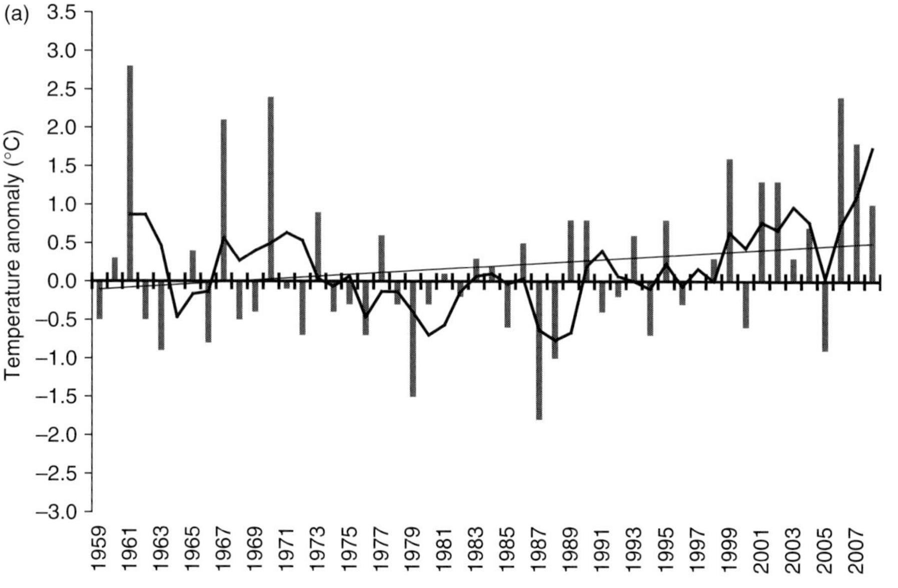

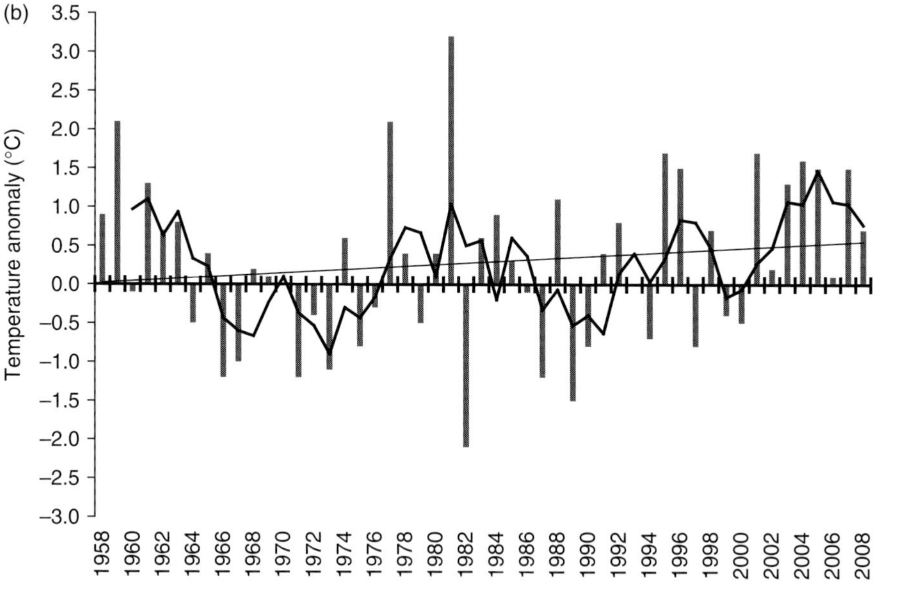

Figure 10.5 (a) Radiosonde record for December, at latitude 67.5°N, 700 hPa altitude. (b) Radiosonde record for January, at latitude 67.5°N, 700 hPa altitude. (c) Radiosonde record for February, at latitude 67.5°N, 700 hPa altitude. (d) Radiosonde record for winter, at latitude 67.5°N, 700 hPa altitude. See text for discussion. (The December data are from the previous year; hence the December plots start one year later than the January and February plots.)

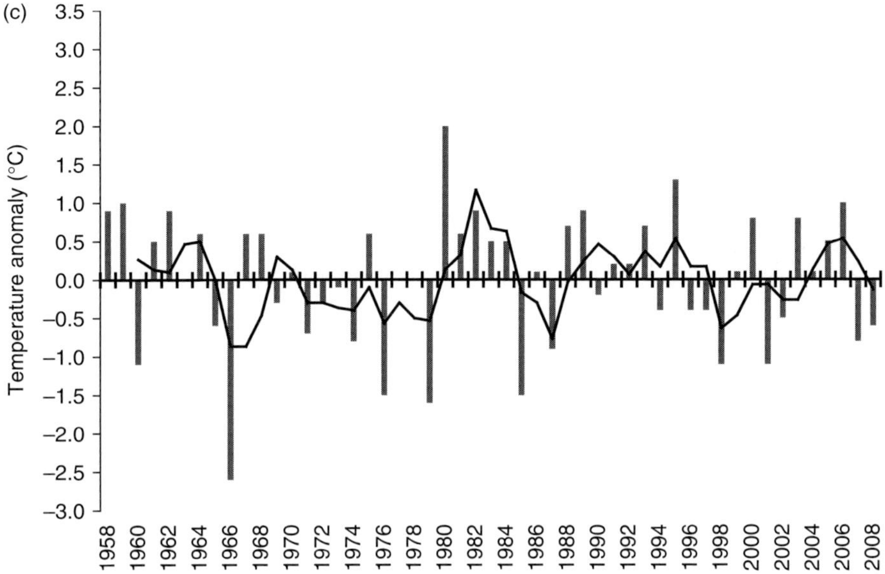

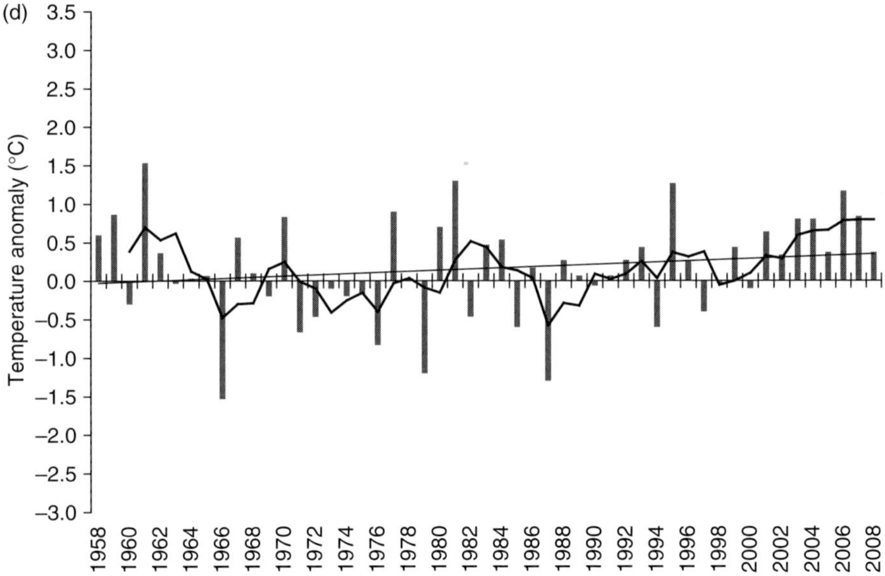

Figure 10.5 (*cont.*)

References

Agudelo, P. A. and Curry, J. A. (2004) Analysis of spatial distribution in tropospheric temperature trends. *Geophysical Research Letters* **31**, L2207.

Angell, J. K. (2003) Effect of exclusion of anomalous tropical stations on temperature trends from a 63-station radiosonde network, and comparison with other analyses. *Journal of Climate* **16**, 2288–95.

Christy, J. R. and Norris, W. B. (2004) What may we conclude about tropospheric temperature trends? *Geophysical Research Letters* **31**, L0621.

Christy, J. R. and Spencer, R. W. (2005) Correcting temperature datasets. *Science* **310**, 927.

DuBois, J. L., Multhauf, R. P. and Ziegler C. A. (2002) *The Invention and Development of the Radiosonde.* Smithsonian Studies in History and Technology, No. 53. Smithsonian Institution Press: Washington DC.

Free, M. and Seidel, D. (2005) Causes of different temperature trends in radiosonde upper-air datasets. *Journal of Geophysical Research* **10**.

Free, M., Durre, I., Agular, E., Seidel, D., Peterson, T. C., Eskridge, R. E., Luers, J. K., Parker, D., Gordon, M., Lanzante, J., Klein, S., Christy, J., Schroeder, S., Soden, B., McMillin, L. M. and Weatherhead, E. (2002) Creating climate reference datasets: CARDS workshop on adjusting radiosonde temperature data for climate monitoring. *Bulletin of the American Meteorological Society* **83**, 891–9.

Fu, Q. and Johanson, C. M. (2005) Satellite-derived vertical dependence of tropical troposphere temperature trends. *Geophysical Research Letters* **32**, L10703.

Grody, N. C., Vinnikov, K. Y., Goldberg, M. D., Sullivan, J. T. and Tarpley, J. D. (2004) Calibration of multi-satellite observations for climatic studies: Microwave Sounding Unit (MSU). *Journal of Geophysical Research* **9**.

IPCC (2007) *Climate Change 2007: The Physical Science Basis.* Contribution of Working Group I to the Fourth Assessment Report of the Intergovernmental Panel on Climate Change. Cambridge University Press.

Haimberger, L. (2005) *Homogenization of Radiosonde Temperature Time Series Using ERA-40 Analysis Feedback Information.* ERA-40 Project Report Series 23, European Centre for Medium Range Weather Forecasts: Reading, UK.

Karl, T. R., Hassol, S. J., Miller, C. D. and Murray, W. L. (eds) (2006) *Temperature Trends in the Lower Atmosphere: Steps for Understanding and Reconciling Differences.* Report by the Climate Change Science Program and Subcommittee on Global Change Research, Washington DC.

Lanzante, J. R., Klein, S. A. and Seidel, D. J. (2003a) Temporal homogenization of monthly radiosonde temperature data. Part I. Methodology. *Journal of Climate* **16**, 224–40.

Lanzante, J. R., Klein, S. A. and Seidel, D. J. (2003b) Temporal homogenization of monthly radiosonde temperature data. Part II. Trends, sensitivities and MSU comparisons. *Journal of Climate* **16**, 241–62.

Mears, C. A., Schabel, M. C. and Wentz, F. J. (2003) A reanalysis of the MSU channel 2 tropospheric temperature record. *Journal of Climate* **16**, 3650–64.

Pettifer, R. (2002) The history of radiosondes. In: *Proceedings of the RMS Specialist Group Meeting, 13 March 2002*, Science Museum: London.

Randel, W. J. and Wu, F. (2006) Biases in stratospheric temperature trends derived from historical radiosonde data. *Journal of Climate* **19**, 2094–104.

Santer, B. D., Thorne, P. W., Haimberger, L., Taylor, K. E., Wigley, T. M. L., Lanzante, J. R., Solomon, S., Free, M., Gleckler, P. J., Jones, P. D., Karl, T. R., Klein, S. A., Mears, C., Nychka, D., Schmidt, G. A., Sherwood, S. C. and Wentz, F. J. (2008)

Consistency of modelled and observed temperature trends in the tropical troposphere. *International Journal of Climatology* **28**(13), 1703–22.

Sherwood, S., Lanzante, J. and Meyer, C. (2005) Radiosonde daytime biases and late-20th century warming. *Science* **309**, 1556–9.

Strangeways, I. C. (2003) *Measuring the Natural Environment* (2nd edn). Cambridge University Press.

Thorne, P. W. (2008) Atmospheric science: the answer is blowing in the wind. *Nature Geoscience* **1**, 347–8.

Thorne, P. W., Parker, D. E., Christy, J. R. and Mears, C. A. (2005) Uncertainties in climate trends: lessons from the upper-air temperature records. *Bulletin of the American Meteorological Society* **86**, 1437–42.

Vinnikov, K. Y., Grody, N. C., Robok, A., Stouffer, R. J., Jones, P. D. and Goldberg, M. D. (2006) Temperature trends at the surface and in the troposphere. *Journal of Geophysical Research* **111**, D03106.

WMO (1987) *WMO International Radiosonde Comparison (UK 1984, USA 1985): Final Report* (Nash, J. and Schmidlin, F. J., eds). Instruments and Observing Methods Report No 30. WMO/TD-No 195: Geneva.

WMO (1991) *WMO International Radiosonde Comparison – Phase III – Dzhambul, USSR, 1989: Final Report* (Ivanov, A. et al., eds). Instruments and Observing Methods Report No 40. WMO/TD-No 451: Geneva.

WMO (1996a) *Guide to Meteorological Instruments and Methods of Observation* (6th edn). WMO No 8. WMO: Geneva.

WMO (1996b) *WMO International Radiosonde Comparison – Phase IV – Tsukuba, Japan, 1993: Final Report* (Yagi, S., Mita, A. and Inoue, N., eds). Instruments and Observing Methods Report No 59. WMO/TD-No 742: Geneva.

11

Future climate measurements

> Monitoring climate change requires a fundamentally different observational approach to that used in weather forecasting (Thorne 2008).

Climate measurements in the future will continue to be made using surface instruments and remote sensing from satellites. Although satellites have the advantage of full global cover, their measurements still lack the precision of surface measurements. They are also unable to make continuous measurements at any one place and their instruments lump together large areas of surface, or volume of atmosphere, in one measurement. The most important and precise measurements will, therefore, continue to be those made *in situ* at the surface, with instruments in contact with what they are measuring. It is with the future of these that this chapter is concerned.

The Global Climate Observing System

In Chapter 7, the WMO's Global Climate Observing System (GCOS) was explained and its many problems noted. In particular, attention was drawn to the well-intentioned, well-thought-through and well-documented plans for improvements to the global instrument network to allow it to become better suited to following long-term climate trends in the future: known as the Ten Monitoring Principles of the GCOS. It was also explained that these plans are making limited progress because Member Countries are not obliged to take any action, and many are ill-equipped – financially, technically and organisationally – to bring about, and to maintain, the necessary changes. It also needs to be emphasised that the GCOS stations were designed and operated throughout the second half of the nineteenth and all of the twentieth century purely for meteorological purposes such as weather forecasting and, more recently, for aviation. As Thorne (2008) rightly concluded, these instruments were never intended for, nor do they well suit (even when improved), climatological research, for the reasons explained in Chapters 3 and 4.

In conclusion, therefore, this final chapter sets out proposals for a new network of precision instruments aimed specifically at monitoring the climate. These are not intended to replace the existing GCOS network, this still being required for its continuing, old and original job of simple meteorology and weather forecasting for the National Meteorological Services. The GCOS stations should, therefore, all continue as they are, quite independently. This will also maintain the continuity of the old records, which can later be adjusted using the new network.

A new instrument network

Number of stations required

Although adding more stations to a dataset improves the accuracy of the temperature anomalies at the regional scale, Jones (1994) found that the effect of such an increase in numbers had a negligible effect on hemispheric and global annual temperature anomaly averages. By using different numbers of existing stations to calculate hemispheric temperature anomalies (see Chapter 8), Jones (1994) found that even as few as 100 stations per hemisphere were sufficient to obtain a reasonable agreement with the results obtained by using a much larger number of stations. This does not apply in the case of regional estimates, where as many stations as possible are required in order to get the best results. Also, over time spans shorter than one year, more stations are again required.

Spatial degrees of freedom and principal components analysis were touched upon in Chapter 8. The 'correlation decay-length' concept can be added to these as an additional means of estimating the minimum number of stations that would give reasonable hemispheric accuracy. Using these techniques, Jones (1994) found that as few as about 100 stations in the NH and 65 in the SH (over the land) gave hemispheric mean annual anomalies very close to the values obtained when using thousands of stations. Even with as few as 43 stations in the NH and 12 in the SH, results were still comparable. It could be argued, therefore, that a new network of stations only needs to amount to a few hundred locations, although much depends on what is required of the measurements (see below).

The figures given above are only for the land, which accounts for 29% of the globe's surface area. More stations will be required to cover the oceans. However, due to the greater uniformity of climate across the oceans, spacing between the stations could be greater, reducing the overall number needed. Taking the above land figures as guides, a first rough guess of, say, 350 stations to cover the whole globe – land and sea – would be of the right order (150 over land and 200 over the oceans).

The cost of tethered buoys (Chapter 5) is high, and so the use of islands as platforms to cover the oceans is proposed. Because manual instruments require operators, only islands that were large enough for human settlement could be used in the past, and this inevitably meant larger islands, with the result that the land could affect the climate being measured. With low-powered, fully automatic stations, any island can be used, even small rocky outcrops. Because the measurements can be transmitted via satellite, there is no need to visit the stations except for corrective maintenance and periodic routine servicing. It would be perfectly feasible to find 200 such sites well spaced around the globe.

Purposes of the new network

The application considered by Jones was the calculation of hemispheric means. But, as noted in Chapter 8, the gross lumping together of anomalies from widely different climates, seasons and hemispheres does not take account of the different conditions in different parts of the world and at different times of the year, making the procedure of questionable validity, even if temperature anomalies are used rather than absolute temperatures. Despite this, however, since one gross figure is deemed necessary by the climate community (and the media), a new network of stations to accomplish this task, with much improved accuracy and reliability, would be worthwhile. If only one global figure is wanted, this can be accomplished with just a few hundred stations.

It has been argued in earlier chapters, however, that the real clues to understanding climate change lie not at the global but at the regional scale. Such a new network of precision stations would enable changes in local climates to be watched at 350 sites with much improved precision, free of doubts about station accuracy, operator competence and the stability of the site. Data from such a new set of instruments would, in addition, allow data from the existing GCOS network to be compared and corrected. The network would also provide valuable ground-truth against which to calibrate and adjust satellite measurements. The new network thus has at least four separate and immediate roles.

Design of the new stations

General concept

Although this book is about temperature, any new station network will need to measure several variables. We will concentrate here on just three: air temperature (over the land and over the sea), sea surface temperature and precipitation. In addition, stations could also measure relative humidity, solar and net radiation (see Chapter 1), wind speed and direction, and barometric pressure. The instruments for measuring all of these (and many others) are described in detail in my book *Measuring the Natural Environment* (Strangeways 2003). The general structure of the stations will be familiar to meteorologists as an automatic weather station (AWS) comprising a mast with sensors deployed from arms (see Figure 4.3b).

Every station must be the same, globally. It is important that a hotchpotch of different stations all of different designs and all with different accuracies is not the outcome. Tenders would need to be put out to companies to quote against a precise station specification, with the production of the latter being the first stage in the process, based on the outline above. Several companies could be employed to undertake the manufacture and installation of the equipment (see the section *Cost and management* below).

Regular inspection of the stations by an international team is essential. In countries where this is not possible for political and/or cultural reasons, stations should not be installed. Complete openness is a basic requirement and the data must also be made available, free of charge, in near-real-time to all countries hosting stations themselves.

Temperature

The sensors

Temperature sensors were described in Chapter 2 and it is only necessary here to say that the most suitable choice for the new network is the platinum resistance thermometer (PRT). These are stable and can be made to a high-precision specification. An accuracy of ±0.01°C should be the aim. Two sensors would be operated side by side to provide back-up in the case of the failure of one, and also to allow cross-comparisons in order to keep a check on calibration drift.

The screens

It will have been made clear in Chapter 3 that the only accurate way to expose any thermometer is in a fan-aspirated screen. Naturally ventilated screens introduce many errors at low wind speeds and under conditions of strong sunlight. For this reason, I am concerned that many AWSs now appear to be in the process of installation, with climate measurements specifically in mind, using the miniature type of screen that is ventilated only by the wind. Such screens are not suitable for climate work, although they are perfectly adequate for general-purpose meteorology and applications that do not require high accuracy. How the ventilating fan will be powered is discussed later.

Sea surface temperature

The same PRTs can be used to measure SST as described above for air temperature, ruggedly protected to defend against rough seas. Their exposure will depend on local conditions, but as the intention is to use small islands for the stations, a structure would be designed to support two identical sensors at a constant depth, allowing for tidal depth changes. This structure would be anchored a short distance from the island to expose it to open-sea conditions.

Precipitation

The most widely used automatic precipitation gauge is the tipping bucket (Strangeways 2003, 2004) and this is the type best suited to remote unattended stations because of its simplicity, reliability, low power requirements and low cost. The method is sometimes criticised for not having good resolution of precipitation intensity, but with a large funnel to collect adequate amounts of water (a minimum of 10 ml per tip), buckets can be made to tip in 0.1 mm or even 0.05 mm steps. Coupled with 'event-mode' logging (recording the date and time of each tip), this would meet most requirements, including climate-change monitoring.

It is well established that the best way to expose any type of rain gauge is with its orifice at ground level. In this way, wind errors, which are the largest source of raingauge error, are reduced virtually to zero. Where practicable, this method of exposure should be used at the new sites. Where rapidly growing vegetation, blowing sand (or other debris) or sea spray is a hazard, an alternative may be necessary. If gauges cannot be operated at ground level, wind shields can be used to reduce the effect of wind, although they are not as effective as ground-level exposure. Another alternative is the use of rain gauges designed to have aerodynamic profiles (Strangeways 2004), which are cheaper. It would be worthwhile operating both ground-level and aerodynamic or shielded gauges at all sites, one backing up the other.

Other variables

As noted earlier, a new network of stations would have to measure variables other than temperature and precipitation. This is not the place to elaborate further, beyond saying that it would make good sense to include all of the basic variables. This would not increase costs excessively.

Power supplies

Solar panels will be the usual source of power, charging one of the new efficient types of battery such as nickel–metal hydride or the well-proven sealed lead–acid gel battery. Ample solar power is available to operate the sensors, the logger and the data collection platform (which transmits the data to the satellite). I have operated such stations using solar power in Scotland in winter. Wind generators might be appropriate at some locations but, being mechanical, they are more liable to failure.

Power supplies for aspirated screens

There are now many small fans manufactured for cooling the processor in personal desktop computers that use 'brushless', low-voltage, DC motors. These draw only milliamps, making solar power feasible for powering the fans in the aspirated screen, as well as powering the electronics. It would be possible to vary the aspiration rate to suit conditions, with maximum ventilation at times of maximum solar radiation (when power is also plentiful), and reduced, but not stopped, under low-radiation conditions.

Data telemetry

The electronics at the station will include a data logger that interfaces with the sensors, converting their signals to digital form for local logging on-site (as back-up in case of telemetry failure). The measurements will also be transmitted via satellite in near-real-time either through one of the weather satellites or one of the new commercial communication satellite networks such as Iridium. This is now well-established technology (Strangeways 2003) and does not require elaboration here. I operated many such stations via Meteosat as long ago as the 1980s – one from Antarctica – and it is reliable, well-proven technology.

Site locations

On land

The new network should be kept entirely separate from the old conventional, existing sites such as those of the GCOS. These are almost invariably in unsuitable places (for climate research), installed there out of practical convenience because of their need for operators. The new stations should be entirely remote from human habitation, in areas selected to represent the surrounding countryside and unlikely to change over the next century. They should have little or no human contact. They should not be associated with the older stations, in part to avoid the continuation of the old practices associated with early meteorology.

Over the oceans

As described earlier, stations monitoring the ocean climate will be installed on very small uninhabited islands. There are more than enough of these available.

Cost and management

The cost of a station measuring air and sea temperature and precipitation would be around USD 25,000. The network of 350 stations would thus total USD 9 million. Assuming an installation cost of about the same brings the total to USD 18 million. Ongoing operation and maintenance would be extra but modest. The WMO does not fund any of the GCOS stations, these being funded nationally by the NMSs, and is unlikely to be able to fund the new network. In any case, the WMO would not be well suited to managing any new network, due to its slow bureaucracy and conservative nature. Although this way of working suits its traditional tasks well, a new, modern, streamlined group of people able to act quickly, without being tethered to the past, needs to be created for the job of planning, organising and operating the new network.

I will not suggest means of funding. That will need to be discussed. But it could all be done for the cost of one wind farm.

Concluding remarks

From some of the best datasets currently available it is clear that the situation regarding temperature change is complex and not reducible to one single graph or to one simple global truth. One figure simply hides all the details. For clues as to how, and why, the climate is changing, we need to look at the seasons individually, at each specific climate, at different latitudes, at the land and sea separately, and at one hemisphere at a time.

I am aware that by stressing the complexity of the climate system, the uncertainty of the measurements and the intricacy of what they show, I run the risk of offending the committed environmentalist and possibly encouraging the sceptics. I sit in the middle, unhappy with the often petulant tone of the sceptics but also unable to support the appearance of certainty and simplicity that many climatologists try to convey. We have a long way to go yet.

References

Jones, P. D. (1994) Hemispheric surface air temperature variations: a reanalysis and an update to 1993. *Journal of Climate* **7**, 1794–802.

Strangeways, I. C. (2003) *Measuring the Natural Environment* (2nd edn). Cambridge University Press.

Strangeways, I. C. (2004) Improving precipitation measurement. *International Journal of Climatology* **24**, 1443–60.

Thorne, P. W. (2008) Atmospheric science: the answer is blowing in the wind. *Nature Geoscience* **1**, 347–8.

Appendix A

The gas laws

Boyle's law says that for a fixed mass of gas, kept at constant temperature, the volume is inversely proportional to the pressure. This can be expressed as:

$$P \propto 1/V, \quad \text{or} \quad P.V = Constant, \quad \text{or} \quad P_1.V_1 = P_2.V_2$$

Charles's law says that for a fixed mass of gas, kept at constant pressure, the volume is directly proportional to the temperature measured on the absolute scale (degrees Kelvin), this being expressed as:

$$V \propto T, \quad \text{or} \quad V/T = Constant, \quad or \quad V_1/T_1 = V_2/T_2$$

The Pressure law says that for a fixed mass of gas, kept at constant volume, the pressure is directly proportional to the absolute temperature, expressed as:

$$P \propto T, \quad \text{or} \quad P/T = Constant, \quad \text{or} \quad P_1/T_1 = P_2/T_2$$

The ideal gas equation combines these three laws for a fixed mass of gas, giving:

$$P_1.V_1/T_1 = P_2.V_2/T_2, \quad \text{or} \quad P.V = Constant.T$$

The constant is controlled by the amount of gas present, which can be measured in three ways: its mass, the number of moles of gas, n, or the number of molecules of gas present, N. In the latter case the ideal gas equation can be written

$$P.V = N.k.T$$

where k is the Boltzmann constant.

Appendix B

Relative humidity and dew point

Evaporation

While studying pollen under the microscope in 1827, Robert Brown noticed the continuous vibration of very small particles within the pollen grains (not *of* the pollen grains, as is usually stated). This is now known as *Brownian movement* (or *motion*), and it led to formulation of the kinetic theory of gases. This is central to the process of evaporation.

The molecules of liquid water are much closer together than those of a gas, and are separated from each other by just slightly more than the diameter of one molecule. In such close proximity, the atomic particles strongly attract each other by Van der Vaals forces (electrical attractions between molecules), but this force reduces rapidly as their separation is increased. In water vapour, the spacing between molecules is ten diameters or more, depending on their concentration (or *vapour pressure* – that fraction of the total pressure due to the water vapour alone – measured in any of the usual units of pressure, such as millibars or hectopascals) and the attractive force is then extremely small. To produce water vapour from liquid water, the distance between the molecules has to be increased and, to achieve this, work has to be done against the binding Van der Vaals attraction.

During evaporation (at any temperature) water molecules 'boil-off', due to Brownian movement, at a rate proportional to the absolute temperature. Some molecules, however, by chance due to their random motion in the air, find their way back to the water surface, the numbers doing so depending on their concentration in the air, which is directly related to the VP. Some of the returning molecules bounce off the surface but some re-enter the liquid water. The net rate of evaporation is the difference between the boil-off and re-entry rates. So long as the vapour molecules in the air can move away from the water surface, their concentration immediately adjacent to the water surface remains low and a continual flow of molecules away from the surface into the air occurs, at a rate set by the absolute temperature.

Saturation

If, however, the water molecules cannot diffuse away; for example, if the water is in a sealed container, their number, and so the vapour pressure, increases until the boil-off and re-entry rates are equal. Although the same number of molecules continue to escape from the water surface, the same number also now return to it and evaporation ceases. The air in the container is then said to be *saturated*. At any given temperature this balance occurs at a particular vapour pressure, which is known as the *saturation vapour pressure* (SVP).

At 0°C the SVP is 6.11213 hPa over water. There is an equivalent SVP over ice, this being 6.11154 hPa at 0°C. (SVP values are available from hygrometric tables.)

Cooling moist air (while keeping its pressure constant) increases its relative humidity (RH), and a temperature will be reached when the RH is 100%. Further cooling then causes dew to form. This is known as the *dew-point* temperature. There is a similar temperature, below 0°C, known as the *frost-point* temperature (or *hoar-frost point*).

Other units

There are a number of other ways of expressing the water vapour content of air.

- *Relative humidity* (RH) is the ratio (expressed as a percentage) of VP to SVP. Thus 100% RH indicates saturation. This applies just to the temperature of the sample at the time, so without also knowing the air temperature, the RH is not complete information
- *Mixing ratio* – the ratio of the *mass* of water vapour to the *mass* of the dry air with which it is associated
- *Saturation mixing ratio* – the largest *ratio* of water vapour to air at any one temperature
- *Specific humidity* (or *water vapour content*) – the ratio of the *mass* of water vapour to the *mass* of moist air
- *Vapour concentration* (or *absolute humidity*) – the ratio of the *mass* of water vapour to the *volume* of the moist air with which it is associated, which is in effect the density of the water vapour, hence its alternative name of *vapour density*.

Appendix C

The electromagnetic spectrum

Figure C.1 (opposite page and over) illustrates the electromagnetic spectrum, with the extreme ends (low radio frequencies and gamma rays) omitted to save space and also because these extremes of frequency are not used in remote sensing (RS) from satellites, as yet. This figure brings together information regarding the wavelength, frequency and photon energy, the names of the radiation, the spectrum of the Sun and the Earth's long-wave infrared spectrum (see also Figure 1.1 in Chapter 1, which shows the spectrum of solar radiation outside the atmosphere and at ground level), along with an indication of the types of instrument used to measure the various bands. Approximate indications of the bands of radiation that are absorbed and transmitted by the Earth's atmosphere are also shown. These 'windows' are important, since remote sensing of the surface is only possible from space in the bands that are not absorbed by the atmosphere, in particular the visible and the two infrared windows A and B, where transmittance is high. The bands that are absorbed, however, are also important because they also radiate, and this can be used to obtain profiles of temperature and other variables down through the depth of the atmosphere (see the section *Sounders* in Chapter 10).

The division of radiation into ultraviolet, infrared, microwave, etc., does not, of course, indicate a sudden change in the nature of the radiation at the point where the name changes. Except in the case of visible light, the division is partly for convenience, and various breakpoints will be found in different writings; the ones used here are typical and as good as any other. The division of radar into K, X, C, S, L and P bands, as shown in Figure C.1, is also just one of several possible divisions.

The most-used wavebands

The largest concentration of measurements from satellites is in the visible region (0.4–0.75 μm), with only slightly fewer being made in the near infrared (0.75–1.1 μm). The next most common band is in the thermal infrared window marked B in the transmittance graph (10.5–12.5 μm). Fewer sensors operate in the window marked A (3.5–4.0 μm). In the region between 6 and 7 μm water vapour radiates strongly, and a few satellites (such as Meteosat) make measurements of humidity by sensing this frequency band. There is no atmospheric window in this band, since water vapour not only radiates at these frequencies but inevitably also strongly absorbs. Meteosat (which is typical of its type) measures in the three bands 0.45–1.0 μm (visible and near IR), 5.7–7.1 μm (water vapour IR) and 10.5–12.5 μm (thermal IR).

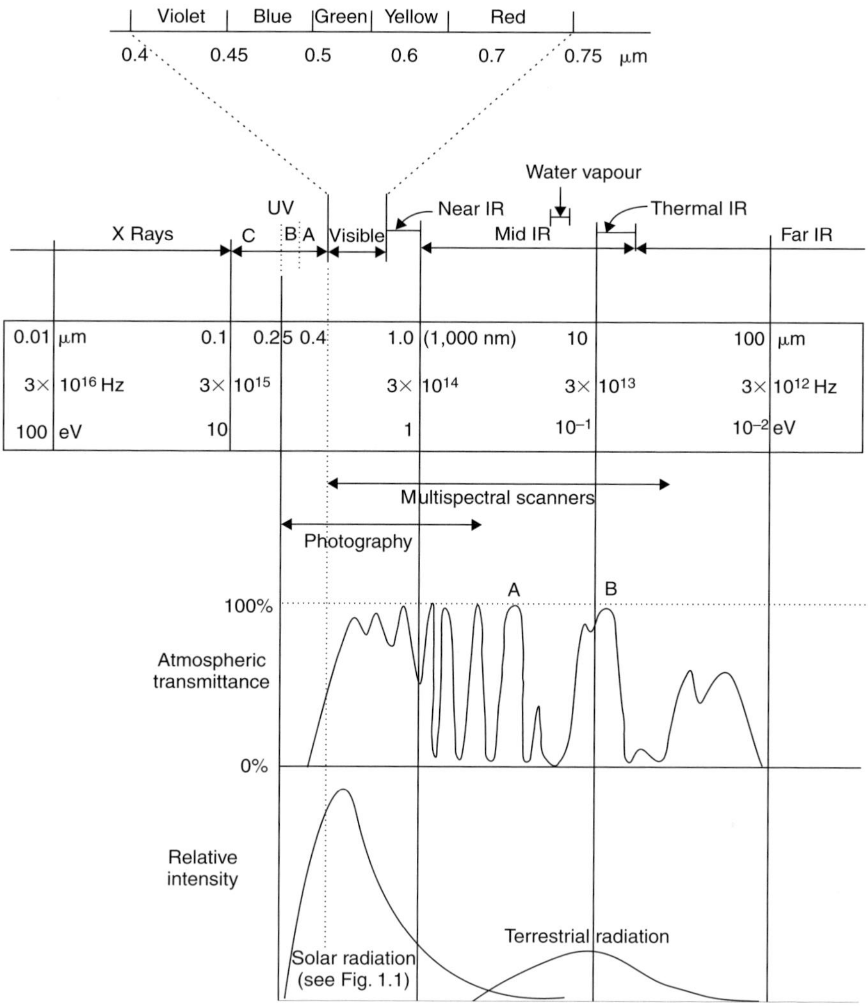

Figure C.1 The electromagnetic spectrum. See text for details.

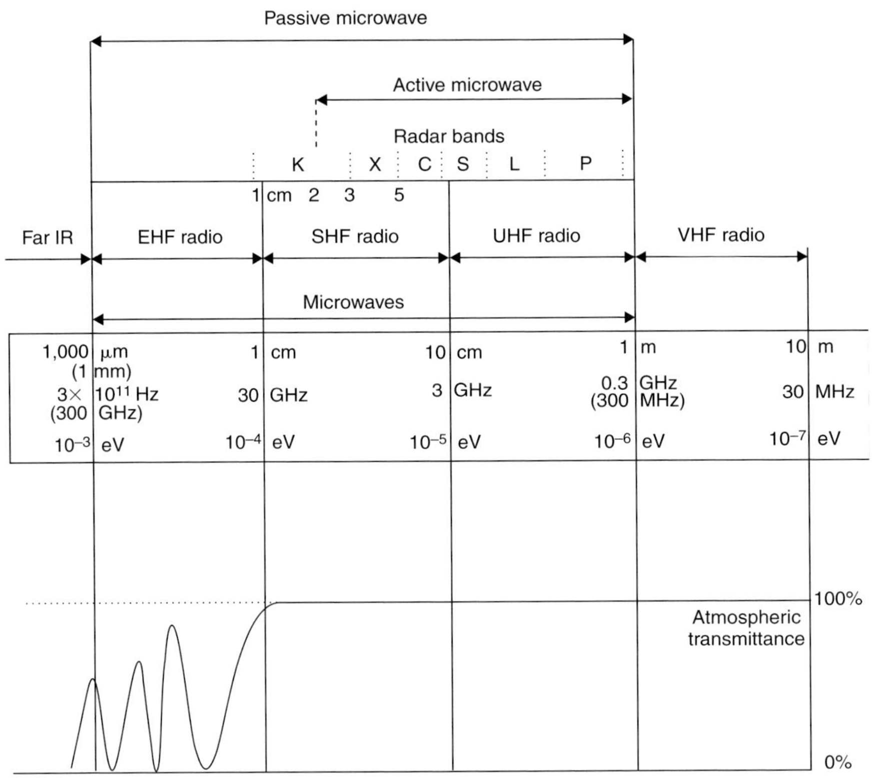

Figure C.1 (*cont.*)

In the microwave region, there are many satellites operating in the C-band using *synthetic aperture radar* (SAR) active sensing techniques, with another group in the X- and K-bands acting as passive scanning radiometers. Temperature is measured by sounders in the 60 GHz region.

Appendix D

Satellite measurements of surface temperature

Satellite orbits

Satellites are operated in two main types of orbit: *polar* and *geostationary*. Atmospheric drag limits orbital height to a minimum of 300 km or more. Satellites at altitudes of 500–2,000 km are usually operated in nearly circular *polar orbits* (giving orbital times of 1–2 hours), 880 km being typical for meteorological satellites. Because the Earth has a slightly greater diameter at the equator than through the poles, a small change occurs in the satellite's orbital period, causing precession of the orbital plane. For a given height, an inclination from the purely north–south path can be calculated that causes the plane to precess one rotation per year. This results in the plane staying fixed relative to the Sun (as the Earth orbits the Sun) and so is termed a Sun-synchronous orbit, the satellite passing over the equator at the same time at each pass through the year. Most *remote-sensing* (RS) satellites are in this type of orbit.

However, the view from polar satellites is constantly changing as they orbit the globe which precludes frequent measurements of the same point on Earth. Some remote-sensing satellites (many fewer), such as Meteosat, are at the much higher altitude of 35,000 km, which causes them to orbit the globe exactly once a day. In such an orbit (around the equator and rotating in the same direction as Earth) satellites appear to be stationary relative to the ground and hence are termed *geostationary* (or more precisely *geosynchronous*) satellites. For more details, and for a comparison of the two orbits, see later under the heading *Spatial and temporal restraints*.

Measuring the radiation

Most sensors on satellites measure the natural radiation from the Sun reflected or scattered from the Earth and its atmosphere and also the long-wave infrared and microwaves originating directly from the Earth and the atmosphere by virtue of their being a 'black body' above absolute zero. Systems measuring radiation from these sources are referred to as passive. It is also possible for an instrument to have its own source of radiation, in particular microwaves (radar) and infrared (generated by laser), on board the platform, with the sensor detecting the reflected radiation. These methods are called active. They do not concern us here.

Radiometers

Visible wavelengths

Visible-light radiometers measure the radiation within a fixed angle of view, either over a wide angle, averaging the whole scene (as in camera exposure meters) or in a highly focused way, using lenses or mirrors for spot measurements. The light is directed onto a sensor such as a photoresistive cell of cadmium sulphide (sensitivity 0.5–0.9 μm) or a silicon photodiode.

Infrared wavelengths

There is also a range of semiconductor photoresistors covering the spectrum from 0.5 to 20 μm, for example indium antimonide, which can be used to detect the shorter IR wavelengths, while mercury cadmium telluride can be used for the longer IR wavelengths in an infrared radiometer.

Scanning radiometers

To combine the high resolution of a visible or IR radiometer with wide areal coverage, it is necessary for the radiometer to scan the view, both vertically and horizontally, as in a TV picture.

Scanning radiometers in polar orbits

If the satellite is moving relative to the ground, as in a polar orbit, the forward (along-track) motion of the satellite provides the X-scan, only the Y-axis having to be scanned by the instrument. If the along-track motion of the satellite is used, the frequency of the across-track scan and the width of the scan strip can be chosen so as to take account of the altitude and speed of the platform, to ensure that adjacent individual strips neither overlap each other nor have gaps between them.

The most usual way of scanning the scene is to rotate a mirror at 45° to the horizontal at several hundred rpm by means of a motor, the radiation being reflected into the system's optics, typically a small Cassegrain telescope. When the mirror is pointing downwards it scans the Earth; as it looks upwards it reflects radiation from a warm on-board calibration source and from the low temperature of deep space a few degrees above absolute zero. In some designs the mirror vibrates rather than rotates.

Scanning radiometers in geostationary orbits

In a geostationary satellite such as Meteosat, both the X- and Y-axes have to be scanned by the satellite, because there is no motion of the satellite relative to the ground. To achieve X and Y scanning, the whole satellite rotates on its axis at 100 rpm, the spin creating the X-scan, while a mirror steps down by a very small angle at each rotation, giving the Y-scan. In this way, over 25 minutes, the satellite scans the entire visible globe, producing 2,500 scan lines.

Multispectral scanners

Multispectral scanners (MSSs) combine a scanning radiometer with spectroscopy. Instead of radiation from the Cassegrain telescope being directed to a single radiation sensor, it is passed through several dichroic mirrors or through a mechanically switched filter, the radiation from each being focused onto a separate sensor, sensitive to a particular part of the spectrum.

The best known of the multispectral scanners in polar orbit is the *Advanced Very High Resolution Radiometer* (AVHRR) operated by the NOAA. The satellite flies at an altitude of 833 km, giving a resolution of 1.1 km at the nadir point, covering a swath of 2,600 km width. Five channels are scanned, one in the visible spectrum (0.58–0.68 μm) for daytime cloud mapping, and one in the near IR spectrum (0.725–1.00 μm) for surface water and ice, with three in the thermal IR spectrum (3.55–3.93, 10.30–11.30 and 11.50–12.50 μm) used for sea surface temperature measurements and night-time cloud mapping.

Typical multispectral scanners in geostationary orbit are those aboard Meteosat and GOES (Geostationary Operational Environmental Satellites), the latter being familiar as the *Visible and Infrared Spin-Scan Radiometer* (VISSR) and the *Spinning Enhanced Visible and Infrared Imager* (SEVIRI), making observations in 12 channels from the visible through the near IR to the far IR at 12.5 μm.

Sounders

Another class of instrument aboard satellites is known as *sounders* or *atmospheric sounders*. Their purpose is to obtain measurements of air temperature and water vapour at different heights down through the depth of the atmosphere from the surface to 40 km altitude, by making observations in both the IR and microwave bands at different wavelengths. These are discussed in Chapter 10.

Calibration of radiometers

Visible channels

Radiometers measuring visible wavelengths, such as the AVHRR, are calibrated before launch. This has the disadvantage that any drift occurring over the years of the satellite's lifetime cannot be corrected or even measured. It is probable that future systems will include on-board calibration.

Infrared channels

IR radiometers are calibrated continuously on board the satellite. Two references are scanned at the end of each line, one being a view of space acting as a black body at 3 K, or virtually absolute zero, the second being a warm target aboard the craft, maintained at a known temperature. A linear relationship between the radiance and the sensor output is derived from these two references, with just a small correction required for non-linearity.

Brightness temperature

Satellite images are composed of numerous individual pixels that, when displayed on a screen, give the impression to the eye of a recognisable continuous two-dimensional view of the Earth. To quantify the data, however, in order to derive numerical information, each pixel is treated as a point value (even if they do represent an area as large as 1 km^2 or more).

In the visible spectrum, the measured radiance is converted to *brightness*, as recognised by the eye or to the reflectance of the surface. Readings in the IR or microwave bands are converted to temperature using the concept of *brightness temperature.* An ideal 'black body' absorbs all the radiation falling on it. Real objects, however, only absorb a fraction of the radiation, known as the absorptivity. The efficiency of radiation, the emissivity, of the same body is the same as its absorptivity at the same wavelength. *Radiance* is the power radiated per unit area per steradian per unit wavelength, and is related to the temperature of the body radiating it, at a given wavelength, via a complex equation involving the speed of light and the Planck and Boltzmann constants in what is known as the Planck function. If radiance in a narrow wavelength band is measured, the temperature of the black body that radiated it can be calculated using the Planck function. This is known as the *brightness temperature.*

However, the radiance is affected by many factors other than the temperature of the actual target, including:

- the elevation of the Sun
- the satellite–Sun angle
- the satellite viewing angle
- the transparency of atmosphere through which the radiation must pass
- Sun-shadowing by higher objects
- the shape of the objects giving shades to the reflected light
- variations of the emissivity of the surface and of the atmosphere.

In consequence, simply measuring the radiance does not give the temperature of the target directly. To obtain this, conversions and corrections have to be made.

Algorithms

Many environmental variables need not be expressed as a precise *quantity,* a simple *yes/no* being sufficient. Cloud/no cloud, snow/cloud, sea/land or desert/forest, for example, all fall into this category. But other variables require numerical values to be derived, including temperature and rainfall amount. Such variables are less amenable to remote sensing (RS) because the radiation reflected from, or emitted by, the Earth's surface or the atmosphere and measured by the satellite radiometer is rarely directly related to the variable.

To deal with these problems, intermediate models or *algorithms* have to be developed to interpret the measurements. These algorithms can be quite complex and may not be very precise, or they may only be applicable to a particular location, or to a particular set of circumstances, or to one season.

Temperature of the surface

Skin temperature

In the case of temperature, the radiation measured in the thermal infrared band gives an indication of the temperature of the surface, be it water or land or vegetation, but only of the very outer surface – the top millimetre or so, literally the skin – with no direct way of measuring the temperature even just a few millimetres below the surface.

Land surface air temperatures

In the case of land surfaces, it is the temperature of the very top skin of the soil or plants that is sensed by the radiometers; the temperature just beneath the surface of the ground is unknowable directly from these measurements. Equally, there is no direct way of knowing what the temperature of the air a metre above the ground is (as measured by an *in situ* thermometer in a Stevenson screen), because the ground temperature is almost never the same as that of the air above it and can differ quite considerably at times. These other temperatures have to be inferred indirectly through a knowledge or estimate of other variables, processes or conditions, and the result is inevitably not as accurate as a primary measurement of the temperature at the surface. It is also a measure of the average over a large area, not of one local point, lumping in surfaces of many different temperatures.

Sea surface temperatures

The same problems exist when making sea surface temperature measurements; just the skin temperature is being sensed. The *in situ* sea surface temperature, measured by a ship or a buoy, is that of the water about 1–10 m below the surface – the bulk temperature – rather than the top few millimetres (see Chapter 5). The situation is further complicated by the state of the surface, whether it is placid or wave-covered, for this affects the emissivity, as well as influencing the mixing of the water.

Measurements of SST are usually made using IR scanning radiometers in the 10.5–12.5 μm window (such as by the AVHRR), the first step being to remove any pixels that contain clouds. For this purpose, special algorithms have been developed. It is then necessary to correct the brightness temperature for the attenuating effects of water vapour, which can introduce errors up to 10 degrees. This is tackled in two ways.

The *differing path length* (*multilook*) approach makes measurements of the same sea area at different angles. Because attenuation is proportional to path length, the two measurements can correct for the attenuation. The European Space Agency satellite ERS-1 carries an IR radiometer known as the *Along Track Scanning Radiometer* (ATSR), which has a dual-angle view of the sea, allowing SST measurements to be corrected for atmospheric interference.

In the *split-window* method, corrections for atmospheric interference are made by measuring the SST at different wavelengths, taking advantage of the fact that water vapour absorption is very wavelength-dependent. The difference in attenuation between a pair of wavelengths is proportional to the difference between a second pair. By looking at two pairs of frequencies in the 10–12 μm band, with one frequency being common to both pairs, the attenuation can be calculated and corrected for. Algorithms using this principle have been developed for both day- and night-time measurements. Different day/night algorithms are necessary because the errors in SST measurement from satellites include factors such as heating of the top layers by sunlight, this being among the more problematic of the errors. Other factors, however, include evaporation, rainfall, currents, upwelling and downwelling, and the mixing of the top layers by wind, all affecting the skin temperature. Night-time readings are often preferred because they eliminate the solar heating problem, although they do not help with the other problems. In comparison with ground-truth measurements, the errors of satellite SST measurements are in the order of half a degree in the IR band, while in the microwave band errors vary from one-quarter to three-quarters of a degree. Problems in merging datasets from different sources, such as surface observations and those from satellites, can be minimised by tuning the algorithms so that they produce results consistent with a reference such as buoy observations. But some error will undoubtedly remain. At microwave frequencies the attenuating effects of clouds are smaller and increasing use is being made of measurements from the Tropical Rainfall Measuring Mission Microwave Imager to measure SST. However, the resolution is coarser (tens of kilometres compared with 1–5 km in the IR).

The strengths and weaknesses of remote sensing

Spatial and temporal restraints

Satellites have the advantage that they can monitor the whole globe and access areas where it can be difficult to site surface stations, in particular the oceans, although this is now less of a problem with automatic weather stations and satellite telemetry. But while satellites operating in geostationary orbits can produce repeat images (of the whole disc) every half-hour, their spatial resolution is 2.5×2.5 km. This is only achievable, however, at the sub-satellite point and, as the field of view extends away from the nadir point, the pixels become increasingly larger and elongated, so that at the limits of view the surface is seen at a grazing angle. The polar regions cannot be seen at all.

Polar orbits may be true polar (at 90° to the equator) or skewed to obtain sun-synchronous operation. Satellites in these orbits are low-flying and so collect images with a higher spatial resolution than those from geostationary satellites. But while polar orbits give better spatial resolution than geostationary orbits, the price paid is reduced time coverage. In the case of satellites in true polar orbits, the poles are in view at each pass (14 times a day typically). As the satellite moves from the poles towards the equator, however, overpasses of any given section of ground become increasingly less frequent, and at the equator there may be many hours without an image of any particular area.

Clouds and night

Clouds are also a problem in the visible and the infrared spectra. In mid-latitudes and in the tropics, clouds can obscure the view frequently and for long periods. If the measurements are also made by an infrequently passing polar satellite, opportunities to observe the surface might be weeks apart. With rapidly changing variables, such as temperature, both of these limitations are obviously a major problem.

Scaling

There is an additional problem, referred to as *scaling up*, which concerns the compatibility of a single point measurement made on the ground with the much larger areal average measurement produced by most satellites, perhaps over an area of 25 km^2. There is also the problem of being sure of the precise geographical location of the RS pixel relative to the ground observation. In addition there are questions of temporal scaling-up from instantaneous satellite measurements to *in situ* means, or of ensuring that the RS measurements apply to the same instant that the ground-truth measurement was made, if it is a spot measurement.

Accuracy and ground-truth

Remote sensing (RS) far outperforms ground-based measurements in many applications, particularly in cloud observation and the detection of the extent of ice and snow, but where precise analogue values are needed, as in the case of temperature, accuracy is limited by the power of the algorithm and by the physical complexity that needs to be modelled. However, it is possible to ameliorate both the lack of a direct correlation between the measured radiation and the variable itself, as well as the effects of atmospheric interference on the passage of the radiation from ground to satellite, by incorporating *in situ* ground-truth measurements. These, combined with the RS quantitative data via the algorithm, greatly increase the accuracy of the RS observations. Indeed it is this combination of ground-truth and remotely sensed data that gives some of the best results in RS. Ground-truth combined with RS also allows the precise *in situ* measurements from the ground-based instruments to be extrapolated in order to give better estimates of the variation between point observations, and thus a better areal average.

It is thus not an either/or but a 'both' choice. The combination of *in situ* with remote data is widely used for most of the variables that hydrologists and meteorologists need to measure. Ground-truth is also essential for the verification of algorithms. That NASA has been prepared to invest so much time, resources and expense in soft-landing instruments on other planets to make *in situ* measurements – of the weather on Mars for example – bears out the importance of ground-truth.

Appendix E

Metadata

Metadata are data about data

They include:

- records of instrument recalibrations
- instrument changes – as straight replacements in the case of failure or breakage, or a change to a different type of instrument
- a move to a new site, but retaining the same name
- a change in the times when readings are made
- a change in how daily means are calculated.

Unfortunately such metadata records are rare. Metadata from the Radcliffe Observatory in Oxford are an example of very good records (Wallace 1997) which are detailed right back to the start of the site in 1772. But most stations do not have such good records, or indeed any records at all, and this makes it difficult to apply the right corrections.

Again the point to note carefully is that it is *changes* that matter. If we are attempting to quantify climate change, we must be sure that nothing else changes that could affect the measurements or that, if such changes have occurred, we know what they are and can allow for them.

Reference

Wallace, J. G. (1997) *Meteorological Observations at the Radcliffe Observatory, Oxford: 1815–1995*. Research Paper No 53. School of Geography, University of Oxford.

Appendix F

The Southern Oscillation Index

There are several ways of calculating the Southern Oscillation Index (SOI) (Ropelewski and Jones 1987; Können et al. 1998), the method used by the Australian Weather Service being as follows:

$$SOI = 10\,(P_{diff} - P_{diffav}) \,/\, SD\,(P_{diff})$$

where P_{diff} is the average Tahiti mean sea-level pressure for the month minus the average Darwin mean sea-level pressure for the month; P_{diffav} is the long-term average of P_{diff} and $SD\,(P_{diff})$ is the long-term standard deviation of P_{diff} for the month.

References

Können, G. P., Jones, P. D., Kaltofen, M. H. and Allan, R. J. (1998) Pre-1866 extensions of the Southern oscillation index using early Indonesian and Tahitian meteorological readings. *Journal of Climate* **11**(9), 2325–39.

Ropelewski, C. F. and Jones, P. D. (1987) An extension of the Tahiti–Darwin Southern Oscillation Index. *Monthly Weather Review* **115**, 2161–5.

Index